INVENTORY MANAGEMENT
MANAGEMENT
Non-Classical Views

Industrial Innovation Series

Series Editor

Adedeji B. Badiru
Department of Systems and Engineering Management
Air Force Institute of Technology (AFIT) – Dayton, Ohio

PUBLISHED TITLES

Computational Economic Analysis for Engineering and Industry
Adedeji B. Badiru & Olufemi A. Omitaomu

Conveyors: Applications, Selection, and Integration
Patrick M. McGuire

Knowledge Discovery from Sensor Data
*Auroop R. Ganguly, João Gama, Olufemi A. Omitaomu, Mohamed Medhat Gaber,
and Ranga Raju Vatsavai*

Handbook of Industrial and Systems Engineering
Adedeji B. Badiru

Handbook of Military Industrial Engineering
Adedeji B.Badiru & Marlin U. Thomas

Industrial Project Management: Concepts, Tools, and Techniques
Adedeji B. Badiru, Abidemi Badiru, and Adetokunboh Badiru

Inventory Management: Non-Classical Views
Mohamad Y. Jaber

STEP Project Management: Guide for Science, Technology, and Engineering Projects
Adedeji B. Badiru

Systems Thinking: Coping with 21st Century Problems
John Turner Boardman & Brian J. Sauser

Techonomics: The Theory of Industrial Evolution
H. Lee Martin

Triple C Model of Project Management: Communication, Cooperation, Coordination
Adedeji B. Badiru

FORTHCOMING TITLES

Beyond Lean: Elements of a Successful Implementation
Rupy (Rapinder) Sawhney

Essentials of Engineering Leadership and Innovation
Pamela McCauley-Bush & Lesia L. Crumpton-Young

Handbook of Industrial Engineering Calculations and Practice
Adedeji B. Badiru & Olufemi A. Omitaomu

Industrial Control Systems: Mathematical and Statistical Models and Techniques
Adedeji B. Badiru, Oye Ibidapo-Obe, & Babatunde J. Ayeni

Innovations of Kansei Engineering
Mitsuo Nagamachi & Anitawani Mohd Lokmanr

Learning Curves: Theory, Models, and Applications
Mohamad Y. Jaber

Modern Construction: Productive and Lean Practices
Lincoln Harding Forbes

Project Management: Systems, Principles, and Applications
Adedeji B. Badiru

Research Project Management
Adedeji B. Badiru

Social Responsibility: Failure Mode Effects and Analysis
Holly Alison Duckworth & Rosemond Ann Moore

Statistical Techniques for Project Control
Adedeji B. Badiru

Technology Transfer and Commercialization of Environmental Remediation Technology
Mark N. Goltz

INVENTORY MANAGEMENT

Non-Classical Views

Edited by

MOHAMAD Y. JABER

CRC Press
Taylor & Francis Group
Boca Raton London New York

CRC Press is an imprint of the
Taylor & Francis Group, an **informa** business

CRC Press
Taylor & Francis Group
6000 Broken Sound Parkway NW, Suite 300
Boca Raton, FL 33487-2742

First issued in paperback 2019

© 2009 by Taylor & Francis Group, LLC
CRC Press is an imprint of Taylor & Francis Group, an Informa business

No claim to original U.S. Government works

ISBN-13: 978-1-4200-7997-5 (hbk)
ISBN-13: 978-0-367-38532-3 (pbk)

Library of Congress Cataloging-in-Publication Data

Jaber, Mohamad Y.
    Inventory management : non-classical views / Mohamad Y. Jaber.
        p. cm. -- (Industrial innovation ; 11)
    Includes bibliographical references and index.
    ISBN 978-1-4200-7997-5
    1. Inventory control. I. Title. II. Series.

TS160.J33 2009
658.7'87--dc22   2009006450

Visit the Taylor & Francis Web site at
http://www.taylorandfrancis.com

and the CRC Press Web site at
http://www.crcpress.com

Dedication

To the soul of my father, and to my wife and sons

Contents

Preface

Inventory is perhaps the most interesting and most researched topic of production and operations management. Inventory is important since it affects our lives; it is everywhere as household inventory, social inventory, and business inventory. Inventory provides flexibility, but it comes at a cost. Fortunes are tied up in inventory and managing it properly is vital for the success of an enterprise.

The earliest scientific inventory management approach dates back to the second decade of the past century (F. W. Harris. 1913. How many parts to make at once? *Factory: The Magazine of Management* 10(2), 136–152), with its golden era in the 1950s and 1960s. The models and concepts viewed inventory as an asset, one that can be converted to cash. This is the view of the classical school of thought. The classical paradigm aims to minimize the total cost of inventory. As markets become more dynamic and competitive, companies are pressured to remain responsive and efficient. In response to these pressures, companies have been making changes to their production and inventory systems. This includes the introduction of new technologies and philosophies, such as just-in-time (JIT), that advocates inventory is a waste and must be reduced. This is the view of the modern school of thought. Despite the differences between the two schools of thought, they both try to answer the two fundamental questions, which are how much to order and when to order. These inventory decisions are usually based on cost parameters that are misleading. Although the classical paradigm, which characterizes both schools of thought, has been successful for years, the question of whether its assumptions are still valid has yet to be answered. In addition, the classical paradigm focused on developing and analyzing inventory models with no regard to the confluence of managerial factors that may affect it directly or indirectly. Furthermore, the emergence of supply chain management and reverse logistics as tools that provide sustainable competitive advantages for companies, and changes in the economy and business activities, require researchers to think outside the classical box of inventory management.

The dire need for a new paradigm has been noted by Professor Attila Chikán, the first vice president and secretary general of the International Society for Inventory Research (ISIR); as well as by other researchers (e.g., Professors M. Bonney and L. G. Sprague). In a recently published paper based on 20 years of ISIR symposium papers, Chikán (2007. The new role of inventories in business: Real world changes and research consequences. *International Journal of Production Economics*, 108(1–2), 54–62) concluded that inventory cannot be managed independently of other company functions, nor can it serve as a buffer between functions and processes, and that cost alone is a misleading performance measure. Recently, the eighth ISIR Summer School organized by Professor Lucio Zavanella and held at the University of Brescia, Italy, had the general theme: "New and Classic Views in Inventory Management:

Advances in Research and Opening Frontiers." The aim was to encourage contributions focusing on emerging trends in inventory studies. The idea to have an edited book on nonclassical views in inventory management, therefore, comes at an appropriate time to feed this emerging interest.

This edited book is unique in collecting emerging works on nonclassical inventory management that feed into the new paradigm. It makes a suitable reference for graduate students and researchers in the areas of inventory theory and management, and supply chain and logistics management.

This book consists of nine chapters.

Chapter 1 introduces and tests a new paradigm of the role of inventories in the operation of business enterprises.

Chapter 2 examines the coverage of inventory issues and concepts in the most popular finance/financial management textbooks.

Chapter 3 examines the relationships between inventory and the environment and discusses how inventory planning could help to alleviate some of these environmental effects.

Chapter 4 aims at showing how energy is related to stocks (and vice versa) and how it may influence them.

Chapter 5 focuses on supply chain management issues from the perspective of the hospital.

Chapter 6 focuses on developing an inventory management strategy for a warehouse supporting a complex emergency relief operation.

Chapter 7 explains how viewing inventories as queues provides a new cause and effect analysis of what creates excessive inventories in industrial systems.

Chapter 8 provides a review of the application of fuzzy set theory in inventory management.

Chapter 9 reviews and discusses the models that use thermodynamic reasoning to model inventory systems.

Mohamad Y. Jaber
Ryerson University

Editor

Mohamad Y. Jaber is a full professor of industrial engineering at Ryerson University, Toronto, Canada. He obtained his Ph.D. in manufacturing and operations management from the University of Nottingham, United Kingdom. His research expertise includes modeling human learning and forgetting curves, workforce flexibility and productivity, inventory management, supply chain management, reverse logistics, and thermodynamic analysis of inventory systems. His research has been supported by the Natural Sciences and Engineering Research Council (NSERC) and the Social Sciences and Humanities Research Council (SSHRC) of Canada. He has published more than 60 articles in internationally refereed journals, including *Applied Mathematical Modeling, Computers & Industrial Engineering, Computers & Operations Research, European Journal of Operational Research, Journal of Operational Research Society, International Journal of Production Economics, International Journal of Production Research,* and *Production Planning & Control.* His industrial experience is in construction management. He is an area editor for *Computers & Industrial Engineering* and is on the editorial boards of the *Journal of Operations and Logistics, Journal of Engineering and Applied Sciences,* and *Research Journal of Applied Sciences.* He continues to serve as a member of the program/advisory committees for the *Annual International Symposium on Supply Chain Management* and the *International Symposium on Logistics.* He is a member of the *European Operations Management Association, Decision Sciences Institute, International Institute of Innovation, Industrial Engineering and Entrepreneurship, International Society for Inventory Research, Production & Operations Management Society,* and *Professional Engineers Ontario.*

Contributors

Benita M. Beamon
University of Washington
Seattle, Washington

Maurice Bonney
University of Nottingham
Nottingham, United Kingdom

Attilla Chikán
Corvinus University of Budapest
Budapest, Hungary

Gerry Frizelle
University of Cambridge
Cambridge, United Kingdom

Alfred L. Guiffrida
Kent State University
Kent, Ohio

Mohamad Y. Jaber
Ryerson University
Toronto, Ontario

Peter Kelle
Louisiana State University
Baton Rouge, Louisiana

Stephen A. Kotleba
University of Washington
Seattle, Washington

Marc J. Sardy
Rollins College
Winter Park, Florida

Helmut Schneider
Louisiana State University
Baton Rouge, Louisiana

Linda G. Sprague
Rollins College
Winter Park, Florida

Sonja Wiley-Patton
Louisiana State University
Baton Rouge, Louisiana

John Woosley
Louisiana State University
Baton Rouge, Louisiana

Simone Zanoni
Università degli Studi di Brescia
Brescia, Italy

Lucio Zavanella
Università degli Studi di Brescia
Brescia, Italy

1 A New Inventory Paradigm
Conceptual Basis and Survey Results*

Attilla Chikán
Corvinus University of Budapest
Budapest, Hungary

CONTENTS

* This article is an integrated, extended, and advanced version of three earlier papers by the author: Chikán (2007), Chikán (2008), and Chikán (2009). Parts of the text of these papers are used here with permission of the editor of *International Journal of Production Economics*.

1.1 OVERVIEW

This chapter introduces and tests a new paradigm of the role of inventories in the operation of business enterprises. It starts out from the fundamental research results of the 1950s, and analyzes the traditional paradigm formulated in this period, called the golden age of inventory research. However, changes in the economic and business environment led to the emergence of the need of a different approach. The proposed new paradigm is introduced and its relation to the traditional one is outlined. Two surveys were conducted among Hungarian managers to test the validity of the new paradigm—one survey among manufacturing managers and another among logistics managers. The results are supportive, showing that managers' thinking mostly corresponds to the logic of the new paradigm. There are a lot of further research opportunities stemming from the paradigm proposition.

1.2 INTRODUCTION

History of researching inventories shows a very interesting pattern. There are peaks and drops both in time and by functional areas, and one can meet extreme standpoints about their importance in the economy and in business.

This chapter puts down the foundations of a new paradigm for researching inventories. We have enough information about the turn of the century changes in enterprise management to put business inventories in a new perspective, getting them out of the passive role assigned to them for decades and including them as an active part of company strategy. The increasing interest in practical inventory management (measured for example by the number of downloads of inventory-related items on the Internet) and in inventory research (see the growing number of participants at inventory conferences and sessions) clearly express the importance of a revival of the best traditions when some of the best minds of economics and management research focused at least part of their interest on inventories.

Our starting point is the golden age of inventory research in the 1950s. We will discuss the paradigm that served as a foundation of research in this era. This traditional paradigm was based on the characteristics of economic and managerial thinking of the time, and served its function very well for decades. However, as the economic and business environment changed, the need for a new paradigm has emerged. This chapter describes the changes in the environment and introduces the elements of a proposed new paradigm.

It is explicitly stated in the exposition of the new paradigm that it has to be validated in practice. Managers' views are tested in two surveys, results of which are summarized in the second part of the chapter. One of the surveys was completed among manufacturing managers, the other one among logistics managers. The survey results support the new paradigm, but further research is necessary for its full elaboration.

1.3 THE TRADITIONAL PARADIGM OF INVENTORY RESEARCH

I think everyone agrees that the golden age of inventory research was in the 1950s. This was the time when both conceptual and mathematical models of inventories were first formulated, following the sometimes interesting but basically very simple approach of earlier decades (Sprague 2000). Whitin (1957) had a truly classic conceptualization of inventory management, but even operations research (OR) hardliners contributed to conceptual clarification (the article most referred to is probably Ackoff [1956]). The fact that some of the best minds in economics and OR joined inventory researchers helped both conceptualization and mathematical formulation of the inventory problem, with the most important contributions coming from Stanford University (Arrow, Karlin, and Scarf 1958; Scarf, Gilford, and Shelley 1963). A summary and analysis of this age is in Girlich and Chikán (2001). A rich collection and analysis of the classical models can be found in Chikán (1990). The most quoted book on managerial applications is Silver and Peterson (1979 and further editions).

It is interesting that economists contributed to development of models that we would now call OR models; very little attention to inventories is attributed to mainstream economics. Perhaps there are understandable reasons for this continuing negligence of inventories in economic research. However, attempting discovery of these reasons is beyond the scope of this chapter. Here I concentrate only on business-level approaches, with reference to some cross-effects between the two study fields.

1.3.1 PILLARS OF THE TRADITIONAL PARADIGM

The traditional paradigm of inventory research is based on three pillars:

1. Inventories can be handled and optimized independently of other managerial circumstances (such as logistics, operations, or financial resources). This assumption made it possible to treat inventories as a single, controlled variable of a feedback control system, where demand is the diverting variable and the order placed is the regulating variable. All are within the inventory system and independent of other fields of company operation. Classical inventory models are mostly based on these control principles. It is worth mentioning that even though models were interpreted mostly on the SKU (stock-keeping unit, i.e., "item") level, this concept can be extended to the company level, handling total inventories of the company as an unstructured volume.

2. The main role of inventories is to serve as a buffer to be used to
   a. smooth business processes, most of all production and sales; and
   b. maintain a flexible connection between the various organizational units (e.g., shops, factories, depots, etc.) of the company.
   The first role is connected to the actual physical processes going on within the company, whereas the second role refers mainly to the managerial (control) processes. These two roles are obviously connected, but they can be clearly distinguished.
3. The performance measure of operation of the inventory system is the level of total cost associated with the sum of holding and replenishing inventories and to handling shortages. In this concept the independency paradigm is reflected, based on the *ceteris paribus* principle. It is assumed that all other things being equal (which assumption can be valid in the case of the subsystems), inventory management can contribute to the increase of profit (the objective of the company as a whole) if it keeps its costs as low as possible. This approach is justified by defining holding cost so that it includes the opportunity cost of inventory investment, for example, by calculating with the average (or, depending on company policy, the marginal) return-on-investment ratio of the company.

1.3.2 EVALUATION OF THE TRADITIONAL PARADIGM

The traditional inventory paradigm was a consequence of some fundamental assumptions about the operation of the business enterprise. These assumptions came primarily from the economics literature. The most important ones are:

- The profit maximizing company, which made the cost-based decisions justified and did not allow the use of noncost measures
- The economies of scale principle, which made use of the independency principle possible by considering inventory decisions as solely volume-dependent and not seeking economies of scope
- The conventional linear organization structure (the existence of functional silos), which again supported the independency principle by making coordination with other functional areas possible and necessary only at the highest level of the hierarchy

The traditional paradigm played a key role in inventory research of several decades after the 1950s. Some of the most important consequences of its application are:

- We learned a lot about item-level inventory systems; by the 1970s we knew practically all we know today about their operation, management, and decision problems.
- There were only few attempts to synthesize this knowledge. Two books of the 1960s must be mentioned: *Analysis of Inventory Systems* (Hadley and Whitin 1963) and *Inventory Systems* (Naddor 1966) are probably the most often cited works. Both are excellent books, but they deal with special classes of inventory models.

- The new management concepts of the period were very negative toward inventories; they handled them only as residua, that is, consequences of managerial decisions focusing on other issues like sales, production, or distribution. Both material requirements planning (MRP) and just-in-time (JIT)—in the extreme—led to the zero inventory illusion or at least to a business atmosphere where inventories were considered as a (not so) necessary evil.

The diffuse character of inventory research and the lack of synthesis are well reflected in the survey made by the International Society for Inventory Research (ISIR; see Chikán [2002a, 2002b]), where respondents were asked to name the most influential papers or books in their research area. There were fourteen items listed by the respondents to this question, and no work was mentioned twice.

A large number of mathematical models was published in this period, but during this time there were very few notable publications about actual managerial issues and business approaches. This led to a situation in the 1970s–1980s when there was a growing and gradually more deeply perceived gap between theory and practice.

Studies of the golden age of inventory research almost totally neglected inventory problems beyond item-level optimization. The ruling paradigm, and especially the assumption that inventory can be managed independently of other functions, did not make it necessary to examine cross-effects of item-level decisions and their relationship with other decisions regarding production, marketing, and so forth; it seemed to be enough to refer to the *ceteris paribus* principle. Classical works of the time did not even refer to the overall inventory problem of the company; adding the item-level optimums was considered to fulfill the job of company-level optimization. However, this common understanding was more or less implicit. The literature barely referred to the potential problems caused by cross-effects of inventories. (The logic used in the OR literature was analogous to the one used by economists who assumed that the model-company's behavior can be extended to the economy as a whole. This logic is reflected in Carzo [1958]). Researchers accepted that item level models express company behavior, however it has not been shown to be an optimal aggregate inventory management rule (Whybark 1994).

It should be noted that the business framework for analyzing the role of inventories was also missing. In the 1970s there were few new results in the "theory of the firm," and even less in something that could be called the "theory of business." However, many of the classical works that we consider fundamental for understanding the general theoretical background of business were published in the period being discussed. Readers may refer to Putterman (1986) and Foss (1997) for a review.

In the same period, several new management concepts and methods were introduced. The most important ones were MRP and JIT, of which at least a side effect should have been reductions in inventories. However, studying inventory data of the period shows that although these methodologies proved successful in many individual company cases, they did not reduce inventories at the level of national economies (Chikán 1994; Bivin 2006). This indirectly suggests that there had been processes going on in the economy since the 1980s–1990s that counterbalanced inventory reduction efforts. This can be considered as a sign of a need for a new business paradigm of inventories, which takes over the old cost-minimizing, inventory-reducing approach.

1.3.3 CONCEPTUALIZATION AND INCORPORATION OF BUSINESS CHANGES BY ACADEMIC RESEARCH

It necessarily took some time before the research community could have conceptualized these changes. There is a special relationship between business practice and research, which is not common in most areas of academic studies. The essence of this relationship is that new, innovative business practices mainly come from actual business life—they precede and not follow research results. The reason for this is that business is under so much pressure to innovate that practitioners are far faster in seeking and introducing new ways of doing things than any other segment of society. The common order of events is:

1. A company introduces a new approach, some innovation that corresponds to the requirement of the changing environment. The need to fit the environment is a fundamental thesis of organization theory (for the contingency approach, see Lawrence and Lorsch [1967]).
2. The new ideas spread rather fast; companies learn from one another in their everyday connections, at business conferences, and so forth. Companies jump on copying the innovations (frequently with the help of consultants, who are among the first to notice new developments and trigger changes). Further developments may follow one of three ways, depending on the actual success of the new approach in practice:
   a. After some time it may turn out that the new idea is a fad, not leading to serious results.
   b. The overall results may be mixed: a new approach can result in good practice at some companies and may prove useless or even harmful at other companies. This is because very frequently the initial application of a new approach is not connected to a prudent analysis of the actual conditions under which a new approach can be successful.
   c. The third case is when a new approach proves to be so successful in such a wide range of cases that business accepts it practically as part of common knowledge.
3. Academic research explains and generalizes empirical research: in case of a fad or a full success it explains why they happened; in case of mixed results it identifies the conditions of successful application.

This approach can be argued. Even though it is supported by long-term observation of business developments, I am sure that there are a lot of exceptions. To illustrate: From among innovations widely known by people dealing with inventories, I would consider BPR in the first, JIT in the second, and MRP in the third category (see a, b, and c above) calling attention again to possible exceptions.

In the aforementioned process, there is an accumulation of knowledge that indicates two things:

a) As the application of new approaches spreads over a given field of management (like marketing, human resources [HR], and inventory management), after some phases, the new approaches may integrate into a new paradigm. A new paradigm for me means a fundamentally new way of thinking about the given management field, which includes different actual principles and practices as well. (The classical reference to paradigm shifts is Kuhn [1996].)

b) Academics have a special role in the process. Compared to businesspeople, they are certainly slow-moving individuals. In business research, academics are more followers than path breakers. However, academics still play a crucial role in business innovation in two respects:

   • They actually explain the developments, help to find the answer to whether a given new idea is a fad or a really useful new idea, and whether the idea is generically beneficial or its usefulness is connected to some special conditions or circumstances.

   • Academic research is indispensable in carrying out the integration of new knowledge into a new paradigm. It does the job of conceptual analysis and generalization; and tests views via empirical studies. Without these two steps no meaningful statement about paradigm changes can be formulated.

1.3.4 EMERGENCE OF THE NEED OF A NEW PARADIGM

As mentioned, the positive effect of the existence of the "old" paradigm was that for the end of the 1960s–beginning the 1970s we knew a lot about the nature and structure of the item-level problems as negative feedback control systems; just in time to give way to a new era that was made necessary by the appearance of MRP systems, with explicit reference to the dependent demand inventories and company-level (or at least product-line-level) optimization. Of course, research on the item-level problems did not disappear, especially since MRP systems actually were in need of models to support lot-sizing decisions.

The appearance and success of MRP systems was mainly a result of the development of information technology and information systems (for the roots and development of MRP systems, see Chikán and Sprague [2008]). MRP was followed by MRP II and distribution requirements planning (DRP), making it even clearer that inventories do not just stand alone but are closely and complexly connected to other company functions. (This idea has already been analyzed by the duly famous Holt et al. book [1960], which was a forerunner of the new approach. A recent exposition of the interfaces between inventories and other functions is given in Stenger [2008].)

The emergence and spread of the JIT concept was a final nail in the coffin of the dominance of the item-level approach in management of inventories. Since JIT simply questioned the fundamental common understanding that inventories are valuable assets of the company, it became explicit that inventories are dependent on the organization of the company as a total entity, and that they can be reduced by the efficient organization of interaction of purchasing, logistics, production, and distribution—practically the whole "real sphere" or "material sphere" of the company. It is common that some authors of that time (e.g., Babbar and Prasad [1998]) practically identified the inventory problem with JIT issues. This seemed to support the practice-based view already known from earlier times that inventories are not managed at all; they are simply residuals of managerial decisions in other fields (mainly in sales and production). There is a lot of anecdotal evidence to support this view, which is 180 degrees different from the classical paradigm of inventory decisions. However, I could not find any research that should have proven that. Instead I believe that Sprague and Wacker (1994) and Whybark (1994) had it right: calling attention to the difference in hierarchy of those who make company-level strategic decisions and those who control actual inventories at a much lower hierarchical level. This duality leads to the appearance of two different approaches: a rather definite inventory reduction approach at the executive level, and a more balanced view at the tactical control level. The former bases its approach mainly on financial return considerations and because inventories, at least at first glance, slow return on investments (since they mean inactive tying up of capital), it is a natural policy requirement to keep them low. However, the interest of low-level managers is completely different. They do not care much about company-level financial return; they are concerned about service levels, production disruptions, shortages of supply, and so forth, and they are actually accountable for these. Therefore, their concept of "optimal" level of inventories is very different than that of their high-level bosses.

Of course there is a trade-off between the two approaches, which can be expressed by balancing the advantages and disadvantages of holding inventories at the two levels described—the strange thing is that there was little, if any, said about that in the inventory literature. Instead there appeared numerous articles and books on "zero inventory," which is a nonsense concept even if the authors in most cases do not mean it literally.

The above reasoning supports the need of a new inventory paradigm that takes into account the built-in duality of inventory management and makes it necessary to examine both levels and their interactions. There must be room in this new paradigm for both quantitative, mathematical models and approaches, and qualitative, organizations-oriented research.

The traditional paradigm provides a sound basis for the former; unfortunately, we have much less historical background for conducting organizational studies of inventory management.

Most companies cannot afford the talent of sophisticated inventory control systems: neither the well-researched and -developed forecasting techniques, nor the performance measures, which would connect the company level and the tactical management level, are affordable at an average company. As Bonney (1994) puts it: "But successful inventory reduction appears likely to be limited to the relatively small number of organizations which can provide the commitment to improving their whole systems." Also, despite the development of information technology

(IT) companies use unsophisticated inventory accounting, which prevents the meaningful connection between the physical and monetary approach to inventories. (I think this is an important component of the macroeconomic "mystery" that interest rates do not influence inventories; see Blinder and Maccini [1991].) Lean inventory management is a nice slogan. Unfortunately, this approach far too often loses sight of the fact that anybody can reduce inventories. The question is the price of this reduction, the often hidden costs of not having inventory on hand. As Sprague and Wacker (1994) put it: "firms do not want inventories 'per se' but want the benefits that inventories can bring." That is why I strongly agree with the practitioner's view: "smart, not lean inventory management will define leaders" (Lawton 2003). It is exactly this smart management that needs the support of a new inventory paradigm. Indeed there are more and more signs of the appearance of this need. This is expressed quite clearly by Miller and Deis (2007). They are consultants with extensive practitioner experience who conclude their paper on aggregate inventory management by saying: "Inventory can be systematically managed." Similarly, Barry (2007) starts his paper with the sentence: "Inventory management and forecasting are strategic issues." These practitioners and consultants clearly indicate the need of a new inventory paradigm. Indeed, the elements of this can be already found in their communication. What is needed is a systematic, scholarly formulation of this paradigm. This is explicitly requested in Invatol (2008): "Despite the many changes that companies go through, the basic principles of Inventory Management and Inventory Control remain the same. Some of the new approaches and techniques are wrapped in new terminology but the underlying principles for accomplishing good inventory management and inventory activities have not changed."

The new paradigm formulated in this chapter may put down the foundations of an appropriate approach. The interest of practitioners can be clearly seen by the huge number (sometimes thousands) of clicks on the inventory management sites of the Internet.

1.4 FUNDAMENTAL CHANGES IN THE GLOBAL ECONOMY

There is very rich literature on changes in the global economy over the past several decades; it is not the purpose of this chapter to discuss this in detail. I only summarize those dimensions of changes that, according to my view, have had the most important influence—directly or indirectly—on inventory management.

1.4.1 TODAY'S ECONOMY

The following characteristics of today's economy are useful for discussion of the new inventory paradigm:

- *Service economy*: The requirements of today's customers are not simply to buy some products and services, but to obtain "solutions" to their problems. This means that companies must offer a combination of products and services.

- *E-economy*: By the use of up-to-date IT, a greater level of integration of the activities of various actors of the economy can be achieved. Also, IT increases the need and availability of speed of action and reaction by actors within the economy.
- *Network economy*: Competition in the market goes on not only among individual actors but also between their networks. Thus, competition and cooperation go on not only in parallel but interlinked, making the whole operation of the economy more complex. Further, network elements and positions are changing constantly, making company relations less predictable and more dynamic. At the same time, this leads to the establishment of long-term relations and strategic alliances as a reaction for safeguarding against impracticable changes.
- *Knowledge-based economy*: Knowledge is now an increasingly important resource, a basis for obtaining competitive advantage. Companies need (and get) more and better-based information and use more sophisticated decision processes.
- *Responsible economy*: Actors within the economy are forced to consider not only their own interests but also those of other stakeholders, including the human and natural environments.
- *Global economy*: Decision makers must consider the whole world economy when making investment, purchasing, or sales decisions, because potential competitors and partners can and will come from anywhere.

The emergence of these characteristics of the economy have led to important changes both in the characteristics of various actors within the economy and (directly or indirectly) in the role of inventories. Since this chapter is focused on the new role of inventories in business, let us now consider changes in business enterprises first and then turn to the new inventory paradigm.

1.4.2 New Characteristics of the Business Enterprise

Observations of real-life business and examination of the literature lead to the conclusion that the purpose and characteristics of the business enterprise have substantially changed in the past few decades. Both declared and actually followed goals seem to be moving beyond simple profit maximization. Fierce competition has caused companies to raise their attention to customers' needs and to define the purpose of the business enterprise as "satisfying customer needs at profit." This does not mean a diminishing importance of profit, but does make life more complex for the enterprise: It has to consider customers' interests with at least the same intensity as its own interests. This means that the company is under a double pressure: It has to create value in one and the same processes for both customers and itself.

I cannot go into full detail here regarding the new objectives and characteristics of the business enterprise. But to understand the new role of inventories, I list those characteristics that I consider most important for this current study. These new characteristics of the business enterprise are:

1. *Focus on competitiveness*, which results in an approach to creating and implementing strategy based on consideration of not what we can do well, but what we can do better than our competitors. This comparative view is essential in both strategic and operational decision making.

2. *Functional integration*, which leads to a different view of the connection among functions (sales, logistics, innovation, production, etc.) of the enterprise. Cross-connections are considered not only in the results but right at the outset of processes. Integrated operation is ensured "in process" by coordination of the various functions at a much lower level of hierarchy than within traditional "silo" systems. This approach is often expressed in both formal and informal organizations, for example, in the form of operational cross-functional teams.

3. *Process orientation*, which means a focus on flows within the company (and beyond its borders), considering complex connections among material, financial, and information flows. The main attention of management is now turning to the process of value creation and all elements of operation are judged by their contribution to this process. The value creation process (sometimes called the value chain; see Porter [1985]) crosses traditional functional boundaries, becoming the main carrier of functional integration.

4. *Network (chain) view*, which means that competition goes on in the global economy, not only between companies but among networks of them. Beyond that, the approach is that we talk about customer value creation as a set of a number of business activities comprising a network, within which organizational boundaries sometimes change rather rapidly (i.e., creation of a product requires given activities, which can be carried out in many different kinds of distribution of work, in different organizational and company forms). This approach reflects the fading boundaries of the firm, and provides a framework for understanding outsourcing and strategic alliances.

These four significant characteristics of today's business enterprise ensure its effective and flexible role in society. This is the main reason why (historically speaking) the ever-increasing needs of people are satisfied to an increasing extent, by business as opposed to self-support or governmental activity.

Let us see now how we can determine the roles of inventories under the aforementioned economic and business conditions.

1.5 NEW ROLES OF INVENTORIES: THE NEW PARADIGM

This section analyzes how the emergence of new economic characteristics has changed the role of inventories and how this role fits into the new enterprise. The new role leads to the need of a new research approach as well that considers the two levels of inventory decisions as described in Section 1.3.4, and provides a framework for their joint integration into the new enterprise.

The main idea is that the classical inventory paradigm must be enriched to fit the new situation: the new paradigm must reflect changes in the business and economic environment. It must be made clear that the new research paradigm for inventories should be nested in practice; it has to be validated against actual managerial behavior.

The essence of the change is that we have to move our perception of inventories from a passive to an active role in company strategy. The new situation does not allow us to view inventories as residua; they have to be considered as an active contributor to a company's success. The definition of the active contributor is that, for this role, strategic decisions must be directly focused on inventories, and inventory policies must influence general strategy formulation and implementation throughout the firm. I believe (and will show some supporting trains of thought) that inventories can have this strategic role—not isolated, of course, but in a complex relationship with other strategic components.

A few remarks must be made before introducing the new paradigm:

1. The traditional paradigm should be made part of the new one since its validity and relevance to item-level decisions can be considered proven.
2. The new paradigm was not born all of a sudden. The path has been prepared by improvement processes within company operations, with the introduction of MRP, JIT, supply chain management (SCM), and so forth; that is, by the development of business activity.
3. The actual role of inventories that corresponds to the new paradigm can and will differ by company, depending on the character of its strategy. However, there is enough ground for generalization to permit fundamental research analyses.

Considering these, I hereby formulate a new inventory research paradigm that I believe corresponds with the current state of real-life economic and business developments.

According to the new paradigm, inventories have strategic importance for companies in three interconnected dimensions: value creation, flexibility, and control.

1.5.1 Inventories as Contributors to Value Creation

If we consider the new concepts of the changed economy, and the purpose and characteristics of the business enterprise, it becomes clear that the emergence of the service economy and, in connection with that, the emphasis on customer satisfaction in company operations lead to a basic question: Can inventories actively contribute to customer value creation? The answer is a definite yes, with reference to some evidence:

- Providing "solutions" to customer needs requires the offering of a mix of physical products and services in practically all business cases. This means a far more complex task than simply deciding on finished goods inventories. The decision on the composition of the "mixture" offered is a key strategic decision—for example, what kind of products and services should be provided by a gas station. There are a number of inventory considerations in such a decision that influence what sorts of customer needs will be met, how, at what level and probability, costs, and so forth.

- In the service economy, the main order-winning factor is often the service provided together with the product—guarantees, maintenance, development support, and so forth. The quality of these "solutions" can hardly be measured by cost parameters only, making some traditional inventory decision rules obsolete.

- In the network economy, a new kind of inventory emerges, which I call *relationship inventory*. This is an element of the total inventory of two companies, being in partnership and not only at arm's length with each other. The level and distribution of this inventory between the two companies depend on the actual relationship of the two companies, their individual and joint interests, strategies, and operational decisions. Characterizing factors of relationship inventories are, therefore, dependent on the features of the partnership itself and not just of the individual companies.

1.5.2 INVENTORIES AS MEANS OF FLEXIBILITY

It has been emphasized already in the traditional paradigm that inventories are buffers for meeting requirements of outside environments and inside organizations. Now this role has been greatly enriched, leading to its far greater importance. Inventories are not simply passive buffers sitting there waiting for usage, but are understood as living elements actively used in pursuing strategic objectives.

- Increased vertical integration and process orientation cause much stronger relationships between subsequent stages both within companies and with their partners. This leads to, for example, closer connections between inventories in various stages of fabrication: they become more dependent on one another and on the policies of customer service. Depending on at which level of fabrication policy-directed stocks are kept, lead times for meeting customer demand can be made shorter or longer.

- The locations of inventories in intercompany supply chains influence the smoothness of relations and can be used for optimizing joint profitability of supplier and vendor companies—for example, by utilizing differences in taxation of various countries or some local savings possibilities in shaping a complex distribution system.

- It is often said that in modern logistics systems, information is traveling instead of materials. However, there are many examples that show that the flexibility and speed of an electronic commerce system can be utilized only if it is supported by high-quality logistics systems, which include appropriately sized and located inventories (the prerequisites of flexible customer service).

- As economies of scope have become a strategic issue, the importance of the mix of inventory positions has increased, closely connected to the horizontal integration of intra- and intercompany activities. More market access decisions must rely on a comprehensive horizontal inventory policy.

1.5.3 INVENTORIES AS MEANS OF CONTROL

Inventories have been considered important indicators of macro- and microeconomic phenomena for a long time. They play a key role as business cycle indicators. Also—less often but importantly—the ratio of input and output inventories in manufacturing is used as an indicator of the relation of overall demand and supply in the market (Chikán 1981; Kornai 1992). In countries or markets where this ratio is low (i.e., where companies generally tend to hold inventories on the output side), economists talk about general oversupply, which goes together with a stronger economic position of buyers (customers). In the opposite case (where the input/output inventory ratio is high), it is easy to sell but hard to buy: high input inventories indicate the fear of not being able to buy goods when needed.

This indicator role of inventories and the possibility of controlling business processes by inventory decisions have been seldom used in the past. Recently we have seen examples under the new economic conditions that these can be expected to increase in number and scope. An interesting means of controlling processes in supply chains are vendor managed inventories (VMIs).

A new feature of business is the growing importance of the CSR (corporate social responsibility) concept. As a part of this development, a new branch of company operation has emerged—reverse logistics, which typically includes keeping inventories of items waiting for remanufacturing, reparation, or recycling. Many models show that reverse logistics activity can be controlled through inventory holding policies.

In general we can summarize that—according to this new view and based on the aforementioned reasoning—inventories may be becoming a more active element of companies' strategic decisions and activities. A summary of the paradigm shift can be seen in Table 1.1.

TABLE 1.1
Comparison of the Traditional and the New Paradigm

Traditional	New
Inventories can be managed independently of other company functions.	Inventories are an integrated part of the value chain in close relationships with other company functions.
Inventories serve as buffers between functions and processes.	Inventories serve as strategic tools in achieving customer satisfaction and profit simultaneously.
Cost is the performance measure.	Performance measures are based on the contribution of inventories to finding better solutions to customer needs than competitors are able to.

1.6 MANAGERS' VIEW OF THE NEW PARADIGM SURVEY RESULTS

To examine the existence of the new paradigm, two surveys were conducted among Hungarian managers.

1.6.1 METHODOLOGY

The first survey was carried out in the winter of 2005–2006 and used the following methodology:

- Nine questions were formulated directly oriented to the managers' perception and approach to inventories.
- These questions were connected to a questionnaire used for surveying manufacturing strategies and practices in the framework of IMSS (International Manufacturing Strategy Survey) and the GMRG (Global Manufacturing Research Group). Both surveys had been conducted several times in many parts of the world (including Hungary) and both contain a number of questions related to inventories.

As a result, there are nine questions directly related to our research focus and a set of complementary questions that could be used in the analysis and discussion.

The survey was originally sent to 245 companies. The companies were asked by phone if they were willing to contribute and in case of a positive answer, a student, appropriately prepared, provided the questionnaire to the companies and explained what was necessary. Fifty-one questionnaires were returned, representing a response rate of 21 percent. The research methodology was repeated with basically the same questions for the fourth time in Hungary, therefore there was no need for a pilot study.

The second survey followed exactly two years later in November and December of 2007. For the development of the questionnaire, the experiences of the survey reported in Chikán (2009) were used for a pilot study.

The questionnaire contained fifteen sets of questions: four on the characteristics of the respondent and his/her company, three on the inventory performance of the company, six on the principles and approaches actually applied in the company, and two on the trends influencing their inventory behavior. The philosophy behind the composition of the questions was to ask the managers about their personal experiences and views, and not about general perceptions. I believe that this is the most reliable approach available in similar empirical surveys: generalization is the job of the researcher and should be based on the addition and integration of individual opinions.

The questionnaire was placed on the Web site of the Hungarian Association of Logistics, Purchasing and Inventory Management. A total of 138 responses were received in a roughly six-week period. The sample was random in the sense that we had not chosen any particular segments of the readers of the Web site. However, because the survey was based on the membership of a professional association, it can be expected that the respondents are more informed and at a higher professional level than the average managers in the country.

TABLE 1.2

Managers' View on the Role of Inventories in Company Operation

	Average Scores	
Question	Manufacturing	Logistics
A. Inventory management is an independent company function; the main inventory decisions have to be made independent of other functional areas.	2.86	2.90
B. Inventories can be efficiently managed only as parts of the supply chain, jointly with other company functions.	6.22	5.70
C. Inventories serve as buffers among various functions and processes of a business unit, protecting against consequences of contradictory interests.	4.38	3.70
D. Inventories are strategic tools of company management serving profitability and customer satisfaction.	4.18	4.82
E. The best performance measure of inventory management is total inventory related costs.	4.16	4.01
F. Performance of inventory management can be measured by its effect on the business unit's competitiveness.	4.67	4.85
G. Inventory considerations play a crucial role in forming product-line policy.	3.63	4.12
H. Inventory considerations play a key role in the preparation of manufacturing programs of subsequent vertically integrated units.	4.66	4.00

1.6.2 MANAGERS' OPINION ABOUT THE ROLE OF INVENTORIES

There was one common set of questions in the two surveys asking managers about their view on the role of inventories. There were the same eight questions in both questionnaires, which had to be answered using a 7-point Likert scale. Table 1.2 contains the results of the two surveys combined. From now on we shall refer to the first survey as "Manufacturing," and to the second as "Logistics," referring to the main functional orientations of the managers involved in the two phases.

For both manufacturing and logistics managers, the same two questions were on top: question B had the highest score, F the second highest. These are the crucial questions regarding the new paradigm, which appears to be accepted by managers. The last place is occupied in both cases by the question most explicitly related to the old paradigm. The difference in the order of the other questions can be explained by the professional orientation of the two sets of respondents. It is obvious that the buffering role (question C) does not disappear; it scores high in both groups. The explicit reference to the strategic role of inventories (question D) is better accepted by logistics managers. The analyses of the two samples show that respondents in the logistics group are higher in the organizational hierarchy than those in the manufacturing group.

A brief discussion of the two surveys follows. The logistics questionnaire focused on the managers' opinion on the environment leading to changes in the role of

TABLE 1.3

Factors Influencing Inventory Efficiency

Reason	Average Score	Orientation
Efforts to reduce costs (a)	5.18	Efficiency
General operational effectiveness of the company (b)	5.10	
Buyer relationship development (c)	4.96	Market
Increase in market competition (d)	4.88	
Supplier relationship development (e)	4.80	
Improved IT systems (f)	4.53	Systems
Logistics system improvement (g)	4.50	
Increased customer requirements (h)	4.65	Responsiveness
Improved management methods (i)	4.40	
Wider product portfolio (j)	4.22	

inventories, whereas the manufacturing survey went into details of the internal issues of the new paradigm.

1.6.3 THE LOGISTICS QUESTIONNAIRE

We'll start with the logistics questionnaire, which deals more with the reasons for the emergence of the new paradigm.

1.6.3.1 Main Reasons behind Inventory Tendencies

Table 1.3 shows the results of a set of questions about the influencing factors of inventory efficiency, in the order of importance attributed by the respondents (respondents evaluated each option on a 7-point Likert scale).

A factor analysis provided practically the same results, connecting aspects a, b, g, h, and i into factor 1; d, j, and f into factor 2; and c and e into factor 3. The order of reasons surprisingly creates four groups of reasons, among which efficiency requirements play the key role leading to higher profitability. Market orientation is in second place, while the two further groups serve as means to achieve efficiency and market goals.

From the point of view of our analysis this means that the average company manager basically agrees with the approach to the influencing factors of inventory management, which was advanced in the previous half of this chapter.

1.6.3.2 Business Trends Influencing Inventories

With the set of questions regarding business trends, it was desired to learn to what extent managers identify with the causes of the changing role of inventories—what they consider as the most important developments influencing inventories. Table 1.4 summarizes the results.

TABLE 1.4

Business Trends Influencing Inventories

Business Trend	Average Score
Focus on competitiveness	5.65
Pressure for complex customer service	5.31
Process view, value chain management	4.87
Internal integration of company functions	4.54
Globalization	4.28
Network economy	4.12
Knowledge economy	3.89
E-economy	3.85
Corporate social responsibility	3.51

If one compares the results with the main trends identified in Chikán (2009), then it can be seen that three of the four factors given there occupy the first three places in Table 1.4. Only one (network economy) received a somewhat smaller average score. This shows a rather good correspondence between the theoretical considerations and the view of the managers.

1.6.3.3 Some Further Insights

We have examined the correlation between the choice of principles (Table 1.2) and the trend scores (Table 1.4). Three methods were used and all of them supported a strong connection between the general "world view" of the managers and the inventory principles applied in their practice.

There was also an attempt to find connections between inventory principles and performance. This branch of analysis led to somewhat disappointing results: No significant correlation was found between the "modernity" of inventory principles and the effectivity of managing inventories. This result may lead to speculation about the actual relevance of the new view on inventories. It may be that since our respondents were mostly top managers, their view may differ from those at the tactical level, which may divert the everyday practice from the principles suggested on top. To determine the real reasons of this feature, further research in the area will follow.

1.6.4 THE MANUFACTURING QUESTIONNAIRE

This section analyzes the results summarized in Table 1.2, mirroring a number of questions in the IMSS–GMRG questionnaire. We would like to identify the characteristics of the respondents who gave different answers to the main inventory questions. Simple correlation coefficients and cross-tables were used for the analysis.

First we describe some general conclusions drawn from the analysis, and then review the inventory questions one by one to see the characteristics of different respondents. Before starting the detailed analysis, it should be said that there is no significant connection between the respondents' manufacturing sectors and the answers; from this point of view our sample is homogeneous.

1.6.4.1 Correlation between Opinions on Inventory Roles

The connection between the responses given to the different inventory questions (from now on they will be referred to as A–H) was examined. The following results were obtained.

From Table 1.5 it can be seen that there are only a few pairs of opinions (6 out of 28) where there are significant correlations. However, these can be quite well interpreted:

A–E: Those who tend to consider inventories as independent functions prob-
    ably consider costs as performance measures. These are the traditional
    thinkers.
D–F: Those who consider inventories as strategic tools (performance mea-
    sured by competitiveness). These are the most dynamic thinkers, assign-
    ing an active role to inventories (characteristic representatives of the new
    paradigm).
D–G and F–G: These correlations are consistent with D–F. Strategy and com-
    petitiveness are supported with active product assortment policy.
F–H and G–H: These combinations connect competitiveness and product pol-
    icy with internal vertical cooperation (consider that these are manufactur-
    ing companies where this concept is rather logical).

Interestingly, questions B and C are not in significant correlation with any other opinions. In the case of B, this can be the consequence of the equally high score given by practically all respondents (see the very low relative standard deviation in Table 1.2). For missing Cs, I have found no explanation.

TABLE 1.5
Correlation Coefficients of Opinions of Inventory Roles

Question	A	B	C	D	E	F	G	H
A	1	−0.123	0.278	0.139	0.307*	−0.106	−0.062	−0.053
B		1	−0.034	0.021	0.078	0.100	−0.126	−0.017
C			1	0.197	−0.160	−0.024	0.211	0.186
D				1	0.069	0.493**	0.344**	0.154
E					1	−0.096	0.054	0.032
F						1	0.508**	0.518**
G							1	0.416*
H								1

*Correlation is significant at the .05 level; ** Correlation is significant at the .01 level.

1.6.4.2   Connection between Performance and Inventory Role Perception

There is only a very loose connection between the performance of the individual companies and the perception of inventory roles of their managers. There was a question in the survey asking about the changes in the last three years of the company's:

- Sales revenue
- Market share
- ROS (return on sales)
- ROI (return on investment)

We have found no significant connection between these variables and any of the opinions on the A–H inventory questions. There is only one exception: the correlation is significant at the .05 level between ROS and question B, which means that companies with higher ROS tend to give a larger score to the statement that inventories can be efficiently managed in the supply chains.

1.6.4.3   Characteristics of Respondents Choosing
Different Roles of Inventories

We analyzed in detail the characteristics of the respondents with the eight different approaches to inventories. The following was obtained for the eight options:

A. *Inventory management is an independent company function.* When analyzing the respondents' characteristics, we found that mainly those managers who gave a higher score to this question, where both external and internal cooperation focuses on the physical rather than the value substance of inventories. These managers attribute a significant role to collaborative planning, consignment, or vendor managed inventory, as well as the physical integration with both the supplier and the buyer. Also, there is a significant correlation between the importance of price in the competitive strategy and giving a high score to A. There are several other answers of these managers that suggest that those who place a higher value on security of operation are more willing to give a higher score to A.

B. *Inventories can be efficiently managed as integrated components of the supply chain.* Not surprisingly, those who gave a higher score to this question are those companies that are actually closely integrated in their supply chain. This conclusion is drawn on the basis of several questions that clarify this relationship in the questionnaire. They feel more competitive than the average regarding their ROS and ROI, but not by their sales revenue and market share. Regarding their strategic development goals, they advance quality and reliability of their products, customization, and flexibility in changing quantity and assortment.

C. *Inventories serve as buffers among various functions and processes.* Giving a higher score to C is significantly correlated with several performance measures, which may mean that better-performing companies put more emphasis on smooth operation than the average companies. These companies, similarly to those in B, consider quality and reliability improvement very important (many of them do not perform well in quality measures). There is a significant correlation also between advancing C and the speed of delivery plus high-quality customer service.

D. *Inventories are strategic tools.* Interestingly there are hardly any meaningful characteristics of the correlation between emphasizing D, the strategic role of inventories, and any other important questions of the questionnaire. It seems that those who consider D important form a very mixed group.

E. *Inventory performance is best measured by costs.* Cost is especially important for those who are competing on prices. These companies have a higher performance measure than the average (which, in fact, shows that the cost reduction approach to inventories is not at all obsolete). They do not put great emphasis on flexibility but want to improve unit production costs and productivity.

F. *Inventory performance is to be measured by contribution to competitiveness.* Interestingly, there is no significant correlation between the F perception of inventories' role and the actual competitiveness of the company. The correlation is significant between emphasizing F and sales revenue and market share, but the connections with ROS and ROI are not significant. This can be interpreted to mean that these companies are more interested in growth than in improving intensive efficiency measures. The companies whose managers are in this group put emphasis on capacity utilization and automation, that is, the physical background of increasing manufacturing performance.

G. *Inventory considerations play a crucial role in forming product-line (assortment) policy.* This statement received the second lowest average score, which means that managers do not really rely on inventories in their "horizontal" policy. However, those who gave a high score to this question, showed the most consistent approach to other questions as well: they gave high scores to all product line (assortment) or economics of scale questions. They want to widen their product scale, increase customization, improve assortment flexibility, and improve product quality—they are all related to the main question.

H. *Inventories in vertically integrated manufacturing units.* This question was more or less customized to the sample of respondents, who were manufacturing managers. On the other hand, while statement G was about horizontal policy, it was logical to put this question about the vertical policy of the company. Interestingly there is a significant negative correlation between placing more emphasis on H and the companies' efficiency in material cost management; perhaps managers hope to improve cost efficiency by better inventory management in vertical integration. Otherwise the significant correlations were found mostly where they could be expected: with product quantity and assortment flexibility, with close connection to customers in following order fulfillment, customization, productivity improvement, and throughput time.

1.7 CONCLUSIONS

We have derived a new paradigm of inventory research from the

- traditional paradigm, which originates from the 1950s and served as an excellent basis for research for decades, and
- the changes in economic and business environment of company operation, experienced progressively in the last two to three decades.

The new paradigm is based on the recognition that there is a gap between the interest of higher- and lower-level managers when handling inventories. The former is interested in the contribution of inventories to the fulfillment of the purpose of the company: meeting customers' needs at a profit. For the latter, inventories serve mostly the requirements for smooth operation and avoiding disturbances. This gap could be bridged under the circumstances of the first part of the last century, when optimizing inventories independently of other company functions on the basis of the *ceteris paribus* principle was acceptable. However, in new developments of the economy and business made company operation far too complex for the traditional handling. This results in the need for a new paradigm.

The new paradigm is based on three components of the new role of inventories in business enterprises:

- Inventories are contributors to value creation.
- Inventories are a means of flexibility.
- Inventories are a means of control.

The validity of the new paradigm as a conceptual construct was tested in two separate surveys: one among manufacturing managers (mostly in middle levels of the hierarchy), and the other among logistics managers (mostly in high levels of the hierarchy). Results of both surveys supported the new paradigm; managers seem to think in accordance with it.

The new paradigm looks promising after the tests. However, limitations in some specificities of the conceptual background and of the surveys make further work necessary on a more detailed and better-founded elaboration of the paradigm. It should be explained, for example, why no correlation could be found between efficiency and the views on inventories in either survey. More detailed questionnaires with a more sophisticated methodology should be applied. Also, other approaches, mainly case studies examining the relationship between inventory decisions at various levels of enterprises, may be useful in strengthening the foundations of the new paradigm.

REFERENCES

Ackoff, R. L. 1956. The development of operations research as a science. *Oper Res* 4(3): 271–273.

Arrow, K. J., S. Karlin, and H. Scarf. 1958. *Studies in the mathematical theory of inventory and production*. Stanford, CA: Stanford University Press.

Babbar, S., and S. Prasad. 1998. International purchasing inventory management and logistics research: An assessment and agenda. *Int J Oper Prod Manag* 18(1):6–36.

Barry, C. 2007. *The best of inventory.* http://multichannelmerchant.com/opsandfulfillment/best_inventory_012007/

Bivin, D. G. 2006. Industry evidence of enhanced production stability since 1984. *Int J Prod Econ* 103(1):438–448.

Blinder, A. S., and L. J. Maccini. 1991. The resurgence of inventory research: What have we learned? *J Econ Surv* 5(4):291–328.

Bonney, M. C. 1994. Trends in inventory management. *Int J Prod Econ* 35(1-3):107–114.

Carzo, R. Jr. 1958. The theory of inventory (mis) management. *Bus Horiz* 1(4):103–110.

Chikán, A. 1981. Market disequilibrium and the volume of stocks. In *The economics and management of inventories*, edited by A. Chikán. Amsterdam: Elsevier.

Chikán, A., ed. 1990. *Inventory models.* Dordrecht: Kluwer Academic Publishers.

Chikán, A. 1994. Some micro explanations of global inventory trends. In *Reflections on firm and national inventories (Workshop on Micro Foundations of Macroeconomic Analysis of Inventories)*, edited by A. Chikán, A. Milne, and L.G. Sprague. Budapest: ISIR.

Chikán, A. 2002a. Inventory research in the last two decades—As reflected in the ISIR Symposia, In *20 years of ISIR*, edited by A. Chikán and A. Kőhegyi, 29–36. Budapest: ISIR.

Chikán, A. 2002b. Status and dynamics of inventory research (Comments on survey results). In *20 years of ISIR*, edited by A. Chikán and A. Kőhegyi, 37–62. Budapest: ISIR.

Chikán, A. 2007. The new role of inventories in business: Real world changes and research consequences. *Int J Prod Econ* 108(1-2):54–62.

Chikán, A. 2008. *Managers' view of a new inventory paradigm.* Paper presented at 15th International Symposium on Inventories, Budapest, Hungary.

Chikán, A. 2009. An empirical analysis of managerial approaches to the role of inventories. *Int J Prod Econ,* 118(1):131–135.

Chikán, A., and L. G. Sprague. 2008. *Operations management fads and fashions: A product life cycle approach to seeking truth.* Paper presented at 3rd World Conference on Production and Operations Management, Tokyo, Japan.

Foss, N. J., ed. 1997. *Resources, firms and strategies.* Oxford: Oxford University Press.

Girlich, H. J., and A. Chikán. 2001. The origins of dynamic inventory modelling under uncertainty (the men, their work and connection with the Stanford Studies). *Int J Prod Econ* 71(1-3):351–363.

Hadley, G., and T. M. Whitin. 1963. *Analysis of inventory systems.* Englewood Cliffs, NJ: Prentice-Hall.

Holt, C., F. Modigliani, J. Muth, and H. Simon. 1960. *Planning production, inventories and work force.* Englewood Cliffs, NJ: Prentice-Hall.

Invatol. 2008. *Inventory management inventory control.* http://www.invatol.com (accessed October 18, 2008).

Kornai. J. 1992. The Socialist System: The Political Economy of Communism Princeton NJ: Princeton University Press.

Kuhn, T. 1996. *The structure of scientific revolutions*, 3rd ed. Chicago: University of Chicago Press.

Lawrence, P., and J. Lorsch. 1967. Differentiation and integration in complex organizations. *Admin Sci Q* 12(1):1–47.

Lawton, J. 2003. *Not your father's inventory strategy.* http://www.eknowtion.org/pdf/Inventory_Strategy.pdf (accessed October 18, 2008).

Miller, G., and P. Deis. 2007. *Aggregate inventory management.* http://www.proaction.net/WhitePaper-AggregateInventoryMgmt.htm (accessed October 18, 2008).

Naddor, E. 1966. *Inventory systems.* New York: Wiley.

Porter, M. E. 1985. *Competitive advantage: Creating and sustaining superior performance.* New York: The Free Press.

Putterman, L., ed. 1986. *The economic nature of the firm.* Cambridge: Cambridge University Press.

Scarf, H. E., D. M. Gilford, and M. W. Shelly. 1963. *Multistage inventory models and techniques.* Stanford, CA: Stanford University Press.

Silver, E. A., and R. Peterson. 1979. *Decision systems for inventory management and production planning.* New York: Wiley.

Sprague, L. G. 2000. Lot sizing principles and practice. *11th Int Symp Inventories,* Budapest, Hungary.

Sprague, L. G., and J. G. Wacker. 1994. Microeconomic underpinnings of macroeconomic analyses of inventories. In *Reflections on firm and national inventories (Workshop on Micro Foundations of Macroeconomic Analysis of Inventories)*, edited by A. Chikán, A. Milne, and L. G. Sprague, 109–123. Budapest: ISIR.

Stenger, A. 2008. *Researching and implementing inventory management systems in commercial firms: 40 years of experience.* Paper presented at 15th International Symposium on Inventories, Budapest, Hungary.

Whitin, T. M. 1957. *The theory of inventory management.* Princeton, NJ: Princeton University Press.

Whybark, D. C. 1994. A management perspective of inventories. In *Reflections on firm and national inventories (Workshop on Micro Foundations of Macroeconomic Analysis of Inventories)*, edited by A. Chikán, A. Milne, and L. G. Sprague, 95–108. Budapest: ISIR.

2 Inventory Management
Some Surprising News about Classical Views on Inventory and Some Nonclassical Responses to Traditional Practice

Linda G. Sprague and Marc J. Sardy
Rollins College
Winter Park, Florida

CONTENTS

2.1 OVERVIEW

Management of an organization's inventories would seem to be at (or at least near) the top of the agenda for a firm's financial managers. On the balance sheet, inventories are close to the top of the list of a company's assets and are often the largest number on the list. (Plant and equipment dominate primarily in capital-intensive industries.) For small- to medium-sized manufacturers (SMMs), inventory is too often a cause of bankruptcy, giving false claim to the notion of inventory as an "asset."

This work finds otherwise. Students of business are likely to receive only a cursory introduction to inventory management during their academic studies, and this most likely only at the microlevel. Examination of the most popular finance/financial management textbooks shows a surprising lack of coverage of inventory issues and concepts, particularly at the strategic level. Hope is expressed that the current emphasis on supply chain management (with its surrounding issues) will help to remedy this dangerous gap in knowledge about a firm's structure and operation.

2.2 INVENTORY: WHAT DOES THE TYPICAL STUDENT BUSINESS MAJOR LEARN?

Business schools' standard undergraduate and MBA/Executive MBA curricula include introductions to the basic themes and topics relevant to the study of business and its operations. The typical curriculum includes required courses in statistics/quantitative methods, financial and managerial accounting, micro- and macroeconomics, marketing, organizational behavior, finance and financial management, operations management, and strategy/strategic management—the basic core of all business school programs. Inventory and its management are most likely to be covered in an operations management course, particularly the micro aspects of inventory control: how much and when to order, safety stocks, and perhaps the details and importance of inventory accuracy.

A typical operations management (OM) course will introduce economic order quantity (EOQ) and safety stocks. All of the best-selling OM textbooks have at least one chapter devoted to inventory—its objectives, definitions, control systems, and order quantities (usually led by derivation of the EOQ). Recent OM textbooks include chapters on supply chain management (SCM), with some emphasizing the role of inventories within these systems. See for example, *Operations Management for Competitive Advantage* (Chase, Jacobs, and Aquilano 2004) and *Operations and Supply Management* (Chase, Jacobs, and Aquilano 2009).

Derivation of the EOQ through the total cost equation is the usual first serious step in development of the student business major's introduction to inventory management and control. This starts with a statement of the numbers needed for such an analysis where:

D = forecast annual demand or usage (see marketing or purchasing)
S = cost of one set-up or order charge (see production supervisor or purchasing)
c = the cost of one item (call accounting)
i = the cost of carrying one item in inventory for one year (call finance)
Q = the order quantity

The total annual cost (TAC) equation is:

$$TAC = \frac{Q \cdot c \cdot i}{2} + \frac{D \cdot S}{Q} \qquad (2.1)$$

Applying the magic of calculus (Harris [1913] referred to this as "higher mathematics"), the EOQ is:

$$Q = \sqrt{\frac{2 \cdot D \cdot S}{c \cdot i}} \qquad (2.2)$$

The first half of the TAC equation is a number that should be of interest to the financial management folks: when summed across all of the inventory items for which this calculation is used, this is the resulting number which will appear as part of the typical third line of the firm's balance sheet—inventory. It is also worth noting that the TAC of any "order policy" ("lot for lot," EOQ, customer order requirement, etc.) can be calculated regardless of how the order size decision is made. The second half of this expression is a number of considerable concern to the shop floor (for firms that do manufacturing and/or assembly), since it is the basis for determining how much of the firm's capacity will be spent on setups or changeovers. In an environment where one-time-per-order costs are incurred (setup or changeovers, shipment fees, quality procedures, etc.), this second part in the equation sums up how much will be spent on costs that occur with each batch produced or purchased.

2.3 A "POLICY VARIABLE"

The firm's operations people are not in charge of any of the four pieces of required data: while the data may be readily available through the firm's ERP system, its actual sources lie elsewhere within the organization. The c is accounting's number for the combined materials/labor/overhead cost incurred and which will be recognized when the item accounting transaction recognizes the transfer to the intended next inventory location. Of particular interest here is i—the cost, as a percentage of the item cost, of carrying an item in inventory.

Technically, the cost of carrying an item in inventory should be reasonably straightforward: it's the sum of any taxes, insurance, storage fees, and so forth incurred when holding an item in stock anywhere in the inventory system. For example, all warehousing costs (including overheads) should be summed and then distributed in some sensible fashion across the entire inventory holding. This, as well as any internal transportation costs, is typically judged to be approximately 5 to 7 percent of the cost of the item per year.

But this relatively low number may be missing the point. At a higher level within the organization, the question is to what other uses could this money have been allocated. In other words, what is the opportunity cost of having the cash tied up in inventory (hardly as "liquid" an asset)? Basically, the higher the percent that the operations people use, the lower will be the resulting inventory positions as the calculations are executed.

Table 2.1 shows recommended values of this percentage over the years. As the source line suggests, these "inventory carrying cost" estimates are from the logistics folks, not the finance people. Typically, a selected percentage will be applied across the board for all inventories in the system, and so is often requested by the

TABLE 2.1

Reported Estimates of Inventory Carrying Charges

Year of Publication	Author	Charge as Percent of Item Cost
1951	Melnitzky	25%
1955	Alford and Bangs	25%
1957	Whitin	25%
1958	Aljian	12%–34%
1962	Ammer	20%–25%
1962	Crook	25%
1973	Heskett et al.	28.70%
1974	Hall	20.40%
1984	Cavinato	25%
1986	Bowersox et al.	20%
1988	Coyle et al.	25%–27%
1990	Johnson and Wood	25%

Source: Lambert, D. M., and Stock, J. R. 1993. *Strategic Logistics Management,* 3rd ed., Homewood, IL: Irwin, p. 366.

information systems group as part of bringing up, for example, a new ERP system. The source may be the "legacy" system or a representative of the operations group.

In operations management courses, we tell students that i—this "carrying charge percentage"—is a policy variable, which is a polite way of saying that the percentage is made up to take into account the fact that "inventory is cash that is not flowing" and we need some easy way to help keep inventory positions down. Since even a medium-sized manufacturing company with a standard information system will calculate several thousand lot sizes every month or so, it is at least some help in keeping inventories low. The numbers shown in Table 2.1, which are in common use in manufacturing organizations, are certainly too high to be normally achievable "hurdle rates" for most companies. Hence, the phrase "policy variable" for these expressions of opportunity costs.

The first known publication describing the EOQ and its derivation appeared in *Factory: The Magazine of Management* in February 1913. The author, Ford W. Harris, said:

> Carrying a large stock means a lot of money tied up and a heavy depreciation. It will here be assumed that a charge of ten percent on stock is fair to cover both interest and depreciation. It is probable that double this would be fairer in many instances. (1913/1990, 947)

Given that overall interest rates at the time were in the low single digits, Harris's recommendation of a 20 percent charge is consistent with the recommended policy variable numbers shown in Table 2.1.

2.4 AN ACCIDENTAL FINDING

While waiting in my coauthor's office while a phone call was in process, I (L. G. Sprague) picked up a copy of Brealey, Myers, and Marcus's *Fundamentals of Corporate Finance* and flipped through to the section on inventory management, finding the following paragraph in the introduction to the "Working Capital Management" chapter:

> The second major short-term asset is inventory. To do business, firms need reserves of raw materials, work in progress, and finished goods. But these inventories can be expensive to store, and they tie up capital. Inventory management involves a trade-off between these costs and benefits. In manufacturing companies, the production manager is most likely to make this judgment without direct input from the financial manager. Therefore, we spend less time on this topic than on the management of the other components of working capital. (Brealey et al. 2007, 543; emphasis added)

So much for our advice to students to "call finance" to set this important value for all inventory decisions.

Pulling several other finance/financial management texts from the shelves revealed an oddity. Other standard finance/financial management texts have similar statements (exact copies for other texts with Brealey as an author). But at least Brealey's text has inventory in the index! From the sample of finance/financial management texts on the shelves, one could conclude that there is little interest in the study of inventories and their strategic management within business schools' curricula—at least within the required finance/financial management courses. This suggested the need for a small research project to see if our classical views on what's going on with respect to the strategic management of inventories within business schools' financial courses are correct.

2.5 INVENTORY: THE ABANDONED ASSET?

The question was: Did our first simple sample of textbooks reflect the overall situation within the academic community's handling of the topic of inventory with the next generations of business school graduates? If so, this could have an impact on where and how inventory and its management belongs in a business school curriculum. Alternatively, should inventory and its management be ignored as a topic for future business practitioners?

In the United States and the United Kingdom there are sources of education about inventories and their management outside the business schools. The American Production and Inventory Control Society (now known as APICS: The Association for Operations Management), and the Institute for Operations Management (IOM) in the United Kingdom both offer certification programs in inventory management through professional seminars and courses through universities and colleges. APICS supports education and testing for CPIM (Certified in Production and Inventory Management) and CSCP (Certified Supply Chain Professional) programs. Education for these is provided through university-based courses, society-offered courses focused specifically on the examinations, and internet-based courses. These programs tend to place more emphasis on inventory control (including information systems issues) than on the financial procedures and processes involved in corporate-level decision making with respect to inventory.

If there is little consideration of how inventories are deployed and financed, and of their financial impact on the firm's overall performance, there will be a poor understanding of the firm's financial structure. Our suspicion is that inventory and its management have been largely ignored in the financial literature—most dangerously in the business textbooks in common use. To investigate this possibility, we examined the situation by analyzing the contents of the best-selling finance/financial management textbooks. Details are described in the Appendix.

2.6 RESULTS

See the Appendix for details of our analysis of inventory coverage in finance/financial management textbooks in the United States. The results support our suspicion that business school graduates in the United States receive little instruction regarding the strategic management of inventories. Since OM faculties tend to focus on inventory control, the overall result is a substantial gap in knowledge about the strategic management of inventories.

Our personal experiences with inventory and its management suggest that there is some truth to the idea that inventory and its management are left to the production managers of the world. This considerable "asset" is in fact controlled item by item and, therefore, lies below the radar screen of the firm's strategic decision makers. This can be described as an inventory control dominant scheme rather than an inventory management dominant system. The equivalent for cash management would be strict emphasis on the necessary minutiae of physical control of the bills and coins with little to no attention to the issues of medium- and long-term flows and locations of cash.

The good news on the horizon could be the rise of supply chain management as a topic being studied in business schools' programs, particularly in OM courses and occasionally in partnership with a marketing course or exercise. It is quickly obvious that when complex systems of inventories are being designed, developed, managed, and controlled, everyone needs to be on board—finance, operations, logistics, marketing, and information systems, to name only the most obvious players. Deployment of millions of dollars of assets in a global marketplace involves serious trade-offs that should not be made "without direct input from the financial manager" (Brealey et al. 2007, 543). Perhaps it is time for operations managers to be recognized as necessary participants in the strategic management of this important asset along with their colleagues in marketing and especially finance.

There is a straightforward way to display the relationship between a firm's inventory management and control, and its overall financial health: the Dupont model (see, for example, Skousen et al. 1999, 3–10; Ross, Westerfield, and Jordan 2006, 68–71). This model is more likely to appear in an accounting textbook than a finance/financial management text.

2.7 CONCLUDING REMARKS

For most assets, firms attempt to maximize their holdings to achieve a reasonable balance among the set of assets. Cash balances, investments in marketable securities, plant and equipment, and so forth are all assets whose holdings increase the firm's ability to respond more nimbly to changes in the economic environment. Inventory does not work the same way: it is best kept in motion, moving from suppliers to the firm and on to customers with value being added throughout its entire journey. Too little inventory and the firm may find dissatisfied customers improving the sales of their better-stocked competitors, or an expensive shutdown if manufacturing goes from "lean" to "anorexic." Too much inventory and the firm could drown in a sea of red ink as an economic downturn squeezes the firm between rising payables and short-term debt, shrinking margins and collapsing inventory valuations.

Inventory control is well established as a business field, and education for the basics is generally available through business schools and professional associations. Between required accounting and operations management courses, business students are primarily provided with procedural and operational aspects of inventory control. Strategic management of inventories has not been an important feature in the most widely used finance/financial management textbooks. However, growing interest in supply chains, which inevitably require attention to serious consideration of "place" of product from raw materials through to retail shelves, is beginning to focus on issues that have apparently fallen between the cracks.

There is no need to wrest control of the strategic aspects of inventory management from the finance folks: they appear to have abandoned the field. For the future, we urge those in the expanding supply chain management business to add serious attention to the balance sheet implications of inventory systems and their dynamics (see The Beer Game [http:// beergame.mit.edu] as an attractive and accessible example). Integrating materials from colleagues in systems dynamics should be encouraged.

There are few industrial marketing experts, but this is a field that could be developed within operations management rather than a marketing department, which is typically focused on consumer behavior where the consumed good is not an industrial product. And those of us in operations management should widen our perspective enough to encompass strategic inventory planning and management, perhaps creating links with our marketing and finance colleagues. Inventory—the abandoned asset—needs a new home.

APPENDIX

Attention paid to inventory and inventory management: an investigation of the inventory content of the best-selling finance/financial management business school textbooks in the United States.

Hypothesis: Inventory and its management, as a subject of study, has been largely ignored in the finance/financial management textbooks used in business schools in the United States.

The Database

To establish the degree of support for this assertion, a sample bibliographic data-base of finance/financial management was created from publications contained in the Rollins College library, our personal academic libraries, and the Orange County Public Library, in Orlando Florida. E-books in finance (books available online) were later added to the search. Although this was clearly not a random sample of text-books or of representative libraries, it is our judgment that libraries tend to maintain their textbook holdings based on instructor use and requests. There is no reason for us to conclude that instructor use of textbooks at Rollins is very different from those in use at U.S. business schools, particularly since these are primarily among the most popular in terms of sales in the United States.

When sales representatives from numerous publishing houses were consulted regarding texts most often ordered by finance faculty members, they confirmed that in the past ten years, most instructors in finance did indeed use the most popular textbooks in their courses.

As an example, the following list summarizes the ranking of publisher McGraw-Hill's leading financial management texts for 2007, as supplied by the marketing department. All but two were included in our database and all of the authors listed were represented by at least one textbook in our database. The list gives us reason-able confidence that our database is indeed representative of the universe of finance/financial management textbooks in general use. It is important to note that we are using textbooks as a proxy variable for "finance literature." Trade books and journal articles were not examined, therefore limiting this study. However, textbooks are a significant source of information about the topics for both undergraduate and gradu-ate business students.

U.S. Rankings of Financial Management Textbooks

Undergraduate

1. *Fundamentals of Financial Management* by S. A. Ross, R. W. Westerfield, and B. D. Jordan
2. *Fundamentals of Financial Management, Concise Edition* by E. F. Brigham and J. F. Houston
3. *Essentials of Corporate Finance* by S. A. Ross, R. W. Westerfield, and B. D. Jordan
4. *Foundations of Financial Management* by S. B. Block and G. A. Hirt
5. *Fundamentals of Financial Management* by E. F. Brigham and J. F. Houston
6. *Fundamentals of Corporate Finance* by R. A. Brealey, S. C. Myers, and A. J. Marcus
7. *Principles of Managerial Finance* by L. J. Gitman
8. *Foundations of Finance: The Logic and Practice of Financial Management* by A. Keown, J. D. Martin, J. W. Petty, and D. F. Scott
9. *Principles of Managerial Finance Brief* by L. J. Gitman

Graduate

1. *Financial Management Theory & Practice* by E. F. Brigham and M. C. Ehrhardt
2. *Principles of Corporate Finance* by R. A. Brealey, S. C. Myers, and F. Allen
3. *Fundamentals of Corporate Finance* by S. A. Ross, R. W. Westerfield, and B. D. Jordan

Source: McGraw-Hill.

The next question addressed was which elements from these books should be part of the database. We thought strongly that a key to our study would be the presence or absence of the mention of inventory in each book's index.

There were 91 titles that we judged to be texts in finance/financial management. Our study consisted of the following database elements for each text:

1. Each textbook was given a number.
2. The library call information was recorded unless it was an e-book or collected from the Orange County Public Library in Orlando, Florida.
3. The year of copyright was noted.
4. The title was listed.
5. The author(s) were identified.
6. Whether the word *inventory* appeared in the index (a yes or no response).
7. If it appeared, we recorded the page numbers in which discussion of inventory appeared.
8. The number of pages on which the subject of inventory was covered.
9. In a section labeled "notes" we identified the focus of the discussion (e.g., LIFO [last in, first out], FIFO [first in, first out], valuation, policies, reorder point, etc.)

ANALYSIS

Since most of the database consisted of nominal or qualitative variables, an intricate quantitative statistical analysis of our judgment sample would have been inappropriate. Therefore we used a simple analytical design. A sort was done on a key variable (e.g., notes) and then a sort on year. This gave us the ability to determine whether there had been changes over time in the notes section of our database. To examine some hypotheses we held about the appearance of inventory in the database, the following sorts were examined:

1. A sort by author, then notes
2. A sort by library, then notes, then year
3. A sort by number of pages, then notes, then year
4. A sort by pages, then notes, then year
5. A sort of notes, then number, then year
6. A sort by year, then notes
7. A sort by title, then notes, then year
8. A sort by notes

The texts examined were published between 1960 and 2008. The number of texts by decade before 2000 and by year after 2000 is shown in Table A.2.1. The sample is heavily skewed toward recent books (after the year 2000) because old textbooks tend to be removed from libraries as more recent editions are purchased and shelved. The sort by the number of titles by author disclosed that authors had at most two titles in the database. This occurred eight times, perhaps because older texts were culled in the libraries. This is a conjecture we were not in a position to examine because culled texts do not appear in our database.

TABLE A.2.1
Number of Texts by Decade or Year

Decade	Number of Texts	Year	Number of Texts
1960s	11	2000	13
1970s	7	2001	6
1980s	9	2002	6
1990s	10	2003	7
		2004	5
		2005	10
		2006	3
		2007	2
		2008	2
Total	37		54

A sort of the database by author, then notes indicated the variety of topics covered in the notes. The five categories in our taxonomic scheme are:

1. From a balance sheet viewpoint
   a. Working capital
   b. Current assets
   c. Acid test ratio
   d. Inventory as a fixed asset
   e. Cash flow
   f. Effect on financial statements

2. From a valuation standpoint
   a. Average of cost or market value
   b. LIFO
   c. FIFO
   d. Replacement costs

3. From an inventory costs standpoint
   a. Inventory in process
   b. Finished goods
   c. Turnover
   d. Carrying costs
   e. Ordering costs
   f. Storage costs
   g. Overstocking

4. From a financial management standpoint
   a. Effect on stock volatility
   b. As collateral

    c. Risk

    d. Spreading inventory across the yield curve

    e. Reducing risk

5. From an inventory management standpoint
    a. EOQ for cash, inventory, etc.
    b. MRP (material requirements planning)
    c. JIT (just-in-time)
    d. Inventory levels
    e. Stockout cost
    f. Seasonal fluctuations
    g. GNP (gross national product) effects

The sources of the texts examined in this study are listed in Table A.2.2. The largest number of books came from the personal academic libraries of the author who teaches finance on a regular basis. It is interesting to note that the library does not own as many texts as the author has in his office. The campus library—with both cost and shelving limitations—acquires texts on a limited basis, usually when the books are in use by course instructors. After a year or two with no borrowing, such texts are typically discarded to make space for newer editions.

The number of e-books was a surprise. Usually a computer screen is the media display, and scrolling and highlighting are not necessarily familiar mechanisms associated with texts. E-books are a more recent phenomena. The earliest copyright was 2003; most were copyrighted after 2005. Because these are more recent texts, we speculated that a fairly high percentage would discuss the concept of financial management in relation to inventory. This conjecture was also tested.

Most library copies had Library of Congress call numbers beginning HG4011 ($n = 17$) or HG173, 174 ($n = 7$). Other call numbers were scattered, although they had titles like *The Theory and Practice of International Finance* or *The Theory of Finance*, which were books essentially written for instructional purposes in finance classes.

TABLE A.2.2
Source of Sampled Texts

Source	Number	Percent
Books/e-books	16	17.6
Personal library	45	49.5
Orange County library	2	2.2
Rollins library	28	30.7
Total	91	100

We examined whether the work contained an index. Fifty-four of the texts did (59.3 percent) and thirty-seven did not (40.7 percent). E-books contain no index, although as they are being read on a computer, searches for a term can be easily launched from the keyboard. Of the seventy-three books in the Rollins library and our personal academic libraries, only fifty-four had indices. Texts lacking indices included such finance texts as *Macro Finance*, *Strategic Financial Planning*, and *Barron's Financial Management*. As more e-books become available, it is possible that the index may become a relic of the past.

Eighty percent of the ninety-one sample texts had no index citation for financing inventories or anything having to do with financing inventories. In eleven of the books with index citations, the reference occurred early in the text (i.e., before page 100; $n = 6$) or late in the book (after page 100; $n = 4$). Since the average textbook size was 467 pages, this finding was also a surprise. Placement within a textbook is a function of the intended sequence of the introduction of the text's material. If a citation occurred early in the book it was usually for inventory valuation or the use of inventory as collateral (category 2 in our taxonomy). If the citation occurred later in the book it was usually for category 3 (a cost explanation) or category 5 (an inventory management point of view). Materials at the end of a text are typically deemed of less importance in the context of the whole course; that is, it is material of less importance than material presented earlier so is more likely to be omitted.

When we examined the number of pages devoted to inventories we found fewer than eleven pages in 70 percent of the books. Fewer than 10 percent of the texts devoted twenty-five pages or more to an exploration of inventories as a subject for discussion. This amounted to an average of 5.3 percent of the textbook pages addressed to inventories; hardly what can be called exhaustive treatments. Not a single book devoted fifty or more pages to the topic of inventories and their management. In most of the books where a larger number of pages appeared, the topic of inventories was focused primarily on showing an example of a calculation, such as an inventory valuation problem or an EOQ exercise with cash as the inventory. Table A.2.3 presents the number of books with or without an index, by year.

The thirty-seven books with a copyright before 2000 contained a mention of the topic of inventories eighteen times. The remaining nineteen indices did not include the topic. In texts with a copyright after 1999, there were thirty-six with index citations and eighteen that ignored the topic. However, despite listing inventories in the

TABLE A.2.3

Appearance of an Index Mention of Inventories in the Text by Copyright Year

Index	Before 2000	After 1999	Total
Yes	18	36	54
No	19	18	37
Total	37	54	91

TABLE A.2.4
Notes Categories Mentioned by Years Excluding Unmentioned

Category	Before Year 2000	Percent	Rank	After 1999	Percent	Rank
1. Balance sheet	8	11	6	16	13	4
2. Valuation	2	3	5	15	12	5
3. Costs	20	28	1	35	29	1
4. Financial management	17	24	2	21	17	3
5. Inventory management	13	18	4	28	23	2
6. No mention	11	16	3	7	6	6

Note: Total Mentioned   [as a % of ... 71 mentioned]   [as a % of 122 mentioned]

index, we found that eleven of the books published before 2000 did not appear to discuss anything on the topic in the actual text. Table A.2.4 presents the notes by year and category of mention, the percentage in each category, and the rank of that category (highest being one).

Before the year 2000, seventy different topics were mentioned in the indices, with categories 3 (costs) and 4 (financial management) receiving the most attention and ranking 1 and 2, respectively. Categories 1 (balance sheet) and 2 (valuation) received the least attention, ranking 6 and 5, respectively. In short, costs and financial management were the hot topics, and valuation and balance sheet were the most ignored. After 1999, costs were still hot (ranking first), but now inventory management ranked second, ahead of financial management of inventory, which now ranked third. The previously ignored categories, valuation and balance sheet, were still ranked lowest. The major change in notes was that the omitted category (category 6) was found in 14.3 percent of the books copyrighted before 2000. In those copyrighted after 1999 the number in that category was reduced to 6 percent. It appears that finance books are no longer ignoring inventory as frequently as in the past.

Earlier in this study we indicated we would examine the hypothesis that e-books would ignore the financing of inventories less frequently because they were written more recently. Appendix Table A.2.5 provides data to test this hypothesis.

TABLE A.2.5
Number of Texts with a Citation for Financing Inventory

Type of Text	Citation	Yes	No	Total
E-book		7	9	16
Others		45	30	75
Total		52	39	91

Table A.2.5 shows that e-books did not appear to mention financing inventories significantly more than other books. This was tested by ξ^2 with one degree of freedom.

In general the overall conjecture of this research was that the finance literature seems to give short shrift to the role of inventory and its management. Though we used texts as a proxy for the overall finance literature and a judgment sample of finance books, we feel confident that the empirical evidence presented supports our conjecture, which is focused on what finance/financial management courses in U.S. business schools are covering with respect to inventories and their management.

Our general conclusion is that business school students in the United States learn little about inventories within their required finance/financial management courses. In particular, students learn little to nothing about the role and impact of inventories within complex supply chains. They are likely to have some introduction to EOQ but primarily as a procedure for inventory control. This particular approach fails to emphasize the insight that a doubling of sales is best not accompanied by a doubling of the supporting inventory level; inventory should increase by the square root of the sales growth. Nor is there attention to the fact that when used within fabrication operations, the second half of the TAC equation directly focuses on an element of shop capacity through its inclusion of setup times (a nonproductive use of capacity). Much of this material is covered in most operations management courses and sometimes in a quantitative analysis course that uses the EOQ model as an illustration of effective analyses for inventory control. This leaves a gap in the conveyance of knowledge about inventories and their management. The most important loss is an understanding of the impact on the firm's asset mix from a combination of strategic inventory positioning decisions. This is probably equaled by the impact of a lack of understanding that inventory control systems may be inadvertently driving inventories unnecessarily high because they are focused on detailed control without reference to overall strategic implications.

We had assumed that the required finance/financial management courses in business schools were conveying some sense of the strategic importance of robust inventory decision making at the corporate level. Our analysis suggests that teachers of finance/financial management in business schools—led by the authors of the best-selling textbooks in the field—have abandoned oversight of this critical asset.

REFERENCES

Abu-Mostafa, Y. S. 2000. *Computational finance 1999*. Cambridge: MIT Press.

Agar, C. 2005. *Capital investment & financing: A practical guide to financial evaluation*. Oxford: Butterworth-Heinemann.

Anderson, W. H. L. 1964. *Corporate finance and fixed investment, division of research, graduate school of business administration*. Boston: Harvard University.

Baker, J. C. 1998. *International finance: Management, markets, and institutions*. Upper Saddle River, NJ: Prentice Hall.

Ball, D. A. 2002. *International business: The challenge of global competition*, 8th ed. Boston: McGraw-Hill.

Besley, S., and E. F. Brigham. 2005. *Essentials of managerial finance*, 13th ed. Mason, OH: Thomson South-Western.

Besley, S., and E. F. Brigham. 2008. *Essentials of managerial finance*, 14th ed. Mason, OH: Thomson South-Western.

Bhar, R., and S. Hamori. 2005. *Empirical techniques in finance*. Berlin: Springer Heidelberg.

Bierman, H. 1980. *Strategic financial planning: A manager's guide to improving profit performance*. New York: Free Press.

Block, S. B., and G. A. Hirt. 2005. *Foundations of financial management*, 11th ed. Boston: McGraw-Hill.

Brealey, R. A., and S. C. Myers. 2003. *Principles of corporate finance*, 7th ed. Boston: McGraw-Hill/Irwin.

Brealey, R. A., S. C. Myers, and A. J. Marcus. 2004. *Fundamentals of corporate finance*, 4th ed. Boston: McGraw-Hill/Irwin.

Brealey, R. A., S. C. Myers, and A. J. Marcus. 2007. *Fundamentals of corporate finance*, 5th ed. Boston: McGraw-Hill/Irwin.

Brigham, E. F., and M. C. Ehrhardt. 2002. *Financial management: Theory and practice*, 10th ed. Fort Worth, TX: Harcourt College Publishers.

Brigham, E. F., and M. C. Ehrhardt. 2005. *Financial management: Theory and practice*, 11th ed. Mason, OH: Thomson/South-Western.

Brigham, E. F., and J. F. Houston. 2007. *Fundamentals of financial management*, 11th ed. Mason, OH: Thomson/South-Western.

Bryant, J. W. 1982. *Financial modelling in corporate management*. Chichester, UK: Wiley.

Buljevich, E. C., and Y. S. Park. 1999. *Project financing and the international financial markets*. Boston: Kluwer Academic.

Cain, P. J. 2002. *Hobson and imperialism: Radicalism, new liberalism, and finance 1887–1938*. Oxford: Oxford University Press.

Carrada-Bravo, F. 2003. *Managing global finance in the digital economy*. Westport, CT: Praeger.

Chase, R. B., F. R. Jacobs, and N. J. Aquilano. 2004. *Operations management for competitive advantage*, 10th ed. Boston: McGraw Hill/Irwin.

Chase, R. B., F. R. Jacobs, and N. J. Aquilano. 2009. *Operations and supply management*, 10th ed. Boston: McGraw Hill/Irwin.

Click, R. W., and J. Coval. 2002. *The theory and practice of international financial management*. Upper Saddle River, NJ: Prentice Hall.

Cochrane, J. H., and National Bureau of Economic Research. 2005. *Financial markets and the real economy*. Cambridge: National Bureau of Economic Research.

Collins Compact English Dictionary. 1994. England: Harper-Collins.

Copeland, L. S. 2005. *Exchange rates and international finance*, 4th ed. Harlow, UK: Financial Times Prentice Hall.

Copeland, T. E., Koller, T. and J. Murrin. 2000. *Valuation: Measuring and managing the value of companies*, 3rd ed. New York: Wiley.

Coyle, B. 2000a. *Cash flow control*. AMACOM, Chicago: Glenlake.

Coyle, B. 2000b. *Cash flow forecasting and liquidity*. Chicago: Glenlake.

Coyle, B. 2000c. *Framework for credit risk management*. Chicago: Glenlake.

Coyle, B. 2000d. *Mergers and acquisitions*, library ed. Chicago: Glenlake.

Crouhy, M., D. Galai, and R. Mark. 2001. *Risk management*. New York: McGraw-Hill.

Crum, R. L., E. F. Brigham, and J. F. Houston. 2005. *Fundamentals of international finance*, 1st ed. Mason, OH: Thomson/South-Western.

Damodaran, A. 2001. *Investment valuation: Tools and techniques for determining the value of any asset*, 2nd ed. New York: Wiley.

De Pamphilis, D. M. 2005. *Mergers, acquisitions, and other restructuring activities: An integrated approach to process, tools, cases, and solutions*, 3rd ed. Boston: Elsevier Academic Press.

Downes, J., and J. E. Goodman. 2003a. *Dictionary of finance and investment terms*, 6th ed. Hauppauge: Barron's Educational Series.

Downes, J., and J. E. Goodman. 2003b. *Finance and investment handbook*, 6th ed. Hauppauge, NY: Barron's.

Eiteman, D. K., A. I. Stonehill, M. H. Moffett, and D. K. Eiteman. 2004. *Multinational business finance*, 10th ed. Boston: Addison-Wesley.

Eun, C. S., and B. G. Resnick. 2004. *International financial management*, 3rd ed. Boston: McGraw-Hill/Irwin.

Financial Times Limited. 1998. *The complete finance companion: Mastering finance*. Saddle River, NJ: Prentice Hall.

Ganguin, B., and J. Bilardello. 2005. *Fundamentals of corporate credit analysis*. New York: McGraw-Hill.

Gitman, L. J., and J. Madura. 2001. *Introduction to finance*, 1st ed. Boston: Addison-Wesley.

Goldsmith, R. W. 1969. *Financial structure and development*. New Haven, CT: Yale University Press.

Gough, L. 2002. *Global finance*. Oxford: Capstone.

Graduate School of Management. 1987. *Recent developments in international banking and finance*. Riverside: University of California.

Grabbe, J. O. 1996. *International financial markets*, 3rd ed. Englewood Cliffs, NJ: Prentice Hall.

Gup, B. E. 1980. *Guide to strategic planning*. New York: McGraw-Hill.

Gurley, J. G., and E. S. Shaw. 1960. *Money in a theory of finance*. Washington: Brookings Institution.

Guthmann, H. G., and H. E. Dougall. 1962. *Corporate financial policy*, 4th ed. Englewood Cliffs, NJ: Prentice-Hall.

Harrington, D. R. 2004. *Corporate financial analysis: In a global environment*, 7th ed. Mason, OH: Thomson/South-Western.

Harris, F. W. 1913. How many parts to make at once? *Factory: The Magazine of Management*, 10(2):135–136. (Reprinted in *Operations Research* 38(6):947–950, 1990.)

Helfert, E. A. 1977. *Techniques of financial analysis*, 4th ed. Homewood, IL: R.D. Irwin.

Helfert, E. A., and E. A. Helfert. 2001. *Financial analysis tools and techniques: A guide for managers*. New York: McGraw-Hill.

Husband, W. H., and J. C. Dockeray. 1972. *Modern corporation finance*, 7th ed. Homewood, IL: R.D. Irwin.

Ingersoll, J. E. 1987. *Theory of financial decision making*. Totowa, NJ: Rowan & Littlefield.

Jones, F. J. 1978. *Macro finance: The financial system and the economy*. Cambridge: Winthrop.

Jorion, P. 2001. *Value at risk: The new benchmark for managing financial risk*, 2nd ed. New York: McGraw-Hill.

Kent, R. P. 1969. *Corporate financial management*. Homewood, IL: R.D. Irwin.

Khoury, S. J., and A. Ghosh. 1988. *Recent developments in international banking and finance*. Lexington, MA: Lexington Books.

Lajoux, A. R., and J. F. Weston. 1999. *The art of M&A financing and refinancing: A guide to sources and instruments for external growth*. New York: McGraw-Hill.

Lambert, D. M., and J. R. Stock. 1993. *Strategic logistics management*, 3rd ed. Homewood, IL: R.D. Irwin.

Lee, C. F., J. E. Finnerty, and E. Norton. 1997. *Foundations of financial management*. Minneapolis/St. Paul, MN: West Pub.

Levich, R. M. 2001. *International financial market: Prices and policies*, 2nd ed. Boston: McGraw-Hill/Irwin.

Lindsay, R., and A. W. Sametz. 1967. *Financial management: An analytical approach*, Rev. ed. Homewood, IL: R.D. Irwin.

Madura, J. 2003. *International financial management*, 7th ed. Mason, OH: Thomson/South-Western.

Madura, J. 2006. *International financial management: Abridged*, 8th ed. Eagan, MN: Thomson/ South-Western.

Martin, J. D., S. H. Cox, and R. D. MacMinn. 1988. *The theory of finance: Evidence and applications*. Chicago: Dryden Press.

Mayo, H. B. 1989. *Finance: An introduction,* 3rd ed. Chicago: Dryden Press.

Megginson, W. L., and S. B. Smart. 2006. *Introduction to corporate finance*. Mason, OH: Thomson/South-Western.

Melvin, M. 2000. *International money and finance*, 6th ed. Reading, MA: Addison-Wesley.

Moffett, M. H., A. I. Stonehill, and D. K. Eiteman. 2003. *Fundamentals of multinational finance*. Boston: Addison-Wesley.

Moore, B. J. 1968. *An introduction to the theory of finance: Asset holder behavior under uncertainty*. New York: Free Press.

Moscato, D. R. 1980. *Building financial decision-making models: An introduction to principles and procedures*. New York: Amacom.

Mossin, J. 1972. *Theory of financial markets*. Englewood Cliffs, NJ: Prentice-Hall.

Mumey, G. A. 1969. *Theory of financial structure*. New York: Holt, Rinehart and Winston.

Paxson, D., and D. Wood. 1998. *The Blackwell encyclopedic dictionary of finance*. Malden: Blackwell.

Perry, P. R., and R. A. Brealey. 1996. *Principles of corporate finance*. New York: McGraw-Hill.

Reilly, R. F., and R. P. Schweihs. 1999. *Handbook of advanced business valuation*. New York: McGraw-Hill.

Riahi-Belkaoui, A. 2001. *Evaluating capital projects*. Westport, CT: Quorum.

Ronen, J., and S. Sadan. 1975. *Corporate financial information for government decision making: A research study and report prepared for the financial executives research foundation*. New York: Financial Executives Research Foundation.

Ross, S. A., R. Westerfield, and B. D. Jordan. 2006. *Fundamentals of corporate finance*, 7th ed. Boston: McGraw-Hill/Irwin.

Schwartz, E. 1962. *Corporation finance*. New York: St Martin's Press.

Schwartzman, S. D., and R. E. Ball. 1977. *Elements of financial analysis*. New York: Van Nostrand Reinhold.

Shapiro, A. C. 2003. *Multinational financial management*, 7th ed. New York: Wiley.

Shefrin, H. 2005. *A behavioral approach to asset pricing*. Amsterdam: Elsevier Academic Press.

Shim, J. K., and J. G. Siegel. 1988. *Handbook of financial analysis, forecasting & modeling*. Englewood Cliffs, NJ: Prentice-Hall.

Shim, J. K., and J. G. Siegel. 2000. *Financial management*, 2nd ed. Hauppauge, NY: Barron's.

Skousen, K. F., W. S. Albrecht, J. D. Stice, E. K. Stice, and M. R. Swain. 1999. *Management accounting*, 1st ed. Cincinnati, OH: South-Western.

Standfield, K. 2005. *Intangible finance standards: Advances in fundamental analysis & technical analysis*. Boston: Elsevier Academic Press.

Sutton, T. 2000. *Corporate financial accounting and reporting*. Harlow/Essex, UK: Prentice Hall/Financial Times.

Terry, B. J. 2000. *The international handbook of corporate finance*, 3rd ed. Chicago: Glenlake.

Thygerson, K. J. 1993. *Financial markets and institutions: A managerial approach*. New York: HarperCollins College Publishers.

Van Horne, J. C. 1966. *Foundations for financial management; a book of readings*. Homewood, IL: R.D. Irwin.

Viscione, J. A. 1977. *Financial analysis: Principles and procedures*. Boston: Houghton Mifflin.

Walter, I. 2004. *Mergers and acquisitions in banking and finance: What works, what fails, and why*. Oxford: Oxford University Press.

Weston, J. F., and S. C. Weaver. 2001. *Mergers and acquisitions*. New York: McGraw-Hill.

3 Inventory Planning to Help the Environment

Maurice Bonney
University of Nottingham
Nottingham, United Kingdom

CONTENTS

3.1 INTRODUCTION, CONTEXT, AND PROBLEM STATEMENT

3.1.1 INTRODUCTION

Many people are concerned about the use of carbon-based energy sources, particularly oil, and how this is affecting the world. They are also concerned about population growth, people's lifestyles, and the use and frequent misuse of limited resources by organizations. Creating energy by burning oil, coal, or gas releases greenhouse gases that are thought by most scientists to be a major cause of global warming that, among other things, could melt the polar ice cap, raise sea levels, and increase flooding in many parts of the world. Disputes and wars arising from consequent shortages of land and water will probably increase.

Product design and packaging design are often environmentally poor. This can lead to excessive use of resources in manufacturing and might produce a product that uses too much packaging that ends up as landfill. Unsatisfactory production processes that produce waste and pollution are common.

In the light of energy shortages and the way that financial relationships are changing around the world, most current industrial and business policies will need to be regularly reexamined from a financial, environmental, and human point of view. Given the vast scale of the problems of global warming, resource shortages, and pollution, it is often asked how individuals and organizations can do something that is environmentally useful. The more focused questions that are asked in this chapter are whether and how inventory policies can help alleviate environmental problems. Can we create usable inventory systems that provide a good service but that are environmentally better than current methods?

So far the effects of inventory planning on the environment have been little studied. As a result, much of the content of the chapter is conjectural. The conjectures are based on examining the relationships between inventory and the environment.

Then, how inventory planning could help to alleviate some of these environmental effects is considered. The proposals will be improved with experience in their use.

3.1.2 CONTEXT AND PROBLEM STATEMENT

Inventory is synonymous with stock, which is normally something tangible that can be physically stored. Properly managed stocks can satisfy many needs. They can provide organizational flexibility; they can act as a buffer between different input and output rates; and they can help to overcome or alleviate many problems and situations.

Inventory is generally part of a multiechelon supply chain. The dynamic response of an inventory system to the demand for its stock items is determined by the inventory rules that are used together with the time that it takes to organize, manufacture and transport items so as to replenish the stock. The consequent fluctuations in stocks and their flows, particularly in a multiechelon system, can be surprisingly different from what might be expected intuitively. However, although holding stock can have advantages, there are frequently disadvantages. Understandably, most inventory analyses aim to choose inventory ordering rules that balance these advantages and disadvantages. Cost is the metric that is most used to analyze inventory problems, and traditional inventory analysis aims to minimize the total cost that arises from three sources: the cost of holding stock (surplus cost), the cost of ordering or getting ready to manufacture stock (ordering or setup cost), and the cost of running out of stock (shortage cost). It may be difficult to obtain realistic cost figures in many practical situations.

The potential environmental advantages and disadvantages associated with different inventory policies are examined later in this chapter. Unfortunately, it is usually even more difficult to associate a realistic cost to environmental consequences than it is to assess the financial costs of inventory surpluses, ordering, and shortages that are used by the classical analysis. A realistic environmental analysis, therefore, requires environmental metrics that may be used to assess, analyze, and determine inventory actions. Possible environmental metrics are discussed in Sections 3.3 and 3.4.

The discussion focuses on selecting policies for managing inventory that will accord with the needs of a greener world. Important inventory decisions include choosing which items to store, choosing and applying appropriate ordering rules, deciding where to locate stores, and organizing and operating the stores over the logistics chain. Applying inventory control methods locally to individual stock items or individual stores is unlikely to make a major environmental contribution by itself. However, a much greater contribution is possible if environmentally responsible inventory management is applied to the supply and distribution chains of products. This is because the decisions then include the locations of manufacturing plants and stores, which affect transportation requirements. There is also a need for appropriate transportation and communications infrastructures. Potentially these have important environmental implications, especially when considered within the current context of industrialization and globalization.

Industrialization and globalization are important because they have changed manufacturing locations and increased the distances between where goods are produced and their markets. On the other hand, industrialization and globalization have developed largely on the basis of cost that depends, according to the product, on labor and material costs, the availability of skills, production organization, and production technology. Locations have seldom been chosen because of environmental considerations, even though in many cases, globalization has led to more efficient production and to a redistribution of wealth between nations, both of which may be good. However, problems still arise that are similar to those that occurred during the Industrial Revolution. These problems include pollution and the exploitation of industrial workers. Therefore, the question is not whether globalization is good or bad, but under what conditions is globalization good and when is it less good. Taking account of environmental effects and factors, such as how individuals and communities will be affected, the questions that need asking include:

- What locations should be chosen for manufacturing and for storage?
- How should items be distributed?
- What rules or incentives should be applied to ensure that the production methods used are ethical and that the decisions that are being made are environmentally responsible?

However, even when there are convincing arguments for or against globalization and new specific locations for the stores and manufacturing plants have been proposed and accepted in principle, the changes will require investment and time to plan. Behavior will not change overnight.

Stocks are diverse and will be found in many fields, such as retailing, manufacturing, medical, or military organizations. Within each group, many different products and demands will be found but, because different stocks have different advantages and costs and present different environmental hazards, it is unlikely that the same environmental strategies will be equally applicable to all items. For example, the financial and environmental costs of transporting heavy items are likely to be greater than for lighter items. For these reasons it is desirable to provide some general principles upon which to base environmentally responsible inventory strategies.

3.1.3   The Structure of the Rest of This Chapter

Some general problems have been outlined and it has been suggested that it is important to consider environmental factors when formulating inventory policies.

The rest of this chapter examines whether inventory planning in the broadest sense (i.e., including locating stores and factories, controlling stocks in stores, controlling production levels, etc.) can be made more environmentally responsible. Some general ideas about inventory are discussed in Section 3.2. In Section 3.3 the environment is considered, and then in Section 3.4 inventory and the environment are considered together. Finally, at the end of the chapter, the discussion in Section 3.5 considers some of the wider interaction issues and presents some tentative conclusions and proposals.

3.2 A CONVENTIONAL INVENTORY CONTEXT

3.2.1 REASONS FOR HOLDING AND NOT HOLDING INVENTORY

A useful operational statement of the inventory problem is to decide when and how much of each stock item to order. The objective is usually to balance the advantages and the disadvantages of holding that stock. In most inventory situations, these *when* and *how much* decisions are embedded within a system that ensures that they are implemented and the system used. An inventory system thus includes procedures for record keeping, ordering, receiving, storing, issuing, and dispatching. The ordering decisions may be strongly linked to the manufacturing procedures, especially if the system emphasis is on making for stock as opposed to making to order. Inventory also has important financial implications. Inventory systems, therefore, need to link with the budgeting and financial control systems and how much stock should be held as raw material, in process, as finished products, and in transit. These ideas are now elaborated.

The main decisions associated with holding stock are to:

- Determine which items to stock
- Determine where the stock should be held
- Choose the inventory control systems
- Obtain and equip appropriate storage facilities
- Install the proposed inventory control system
- Implement and operate the inventory control systems that include procedures for obtaining and controlling the stock

Specific operational activities that exist within the chosen inventory control system are to:

- Record stock transactions
- Maintain appropriate records
- Review stock levels at appropriate times
- Order and receive the stock
- Store the stock securely in appropriate conditions (temperature, humidity, etc.)
- Dispose of the stock by using it, selling it, dispatching it, or scrapping it as appropriate
- Perform the necessary administrative activities such as paying suppliers, invoicing customers, stock checking, and so forth

The factors encouraging stock to be held are:

- Ordering: Placing an order incurs costs (including the administrative cost of ordering, the physical cost of receiving and accepting items, and possibly the cost of packing and transporting items). Ordering is often a fixed charge that may be spread over the quantity of items purchased. Thus, the order cost per unit purchased reduces as the quantity purchased increases.

Also, there may be discounts for buying items in larger quantities, further reducing the unit cost. Similarly, unit manufacturing costs are reduced if items are made in quantity and the setup cost per batch is spread over each unit of the complete batch. The unit cost of an item, therefore, usually reduces as the purchase or manufacturing quantity increases; this encourages buying in larger quantities and thus having higher average stock.

- Stock availability provides flexibility because items are available for immediate use rather than needing to wait for an order to be made and for a delivery to arrive. Hence, availability provides flexibility to a company, and reduces the occurrence of shortages and the consequential organizational problems. This is especially true for long-delivery-time items, because it may take a long time before additional items could become available. Stock may also be held to meet random fluctuations in demand, to smooth seasonal fluctuations in demand and supply, and to allow delivery and issue quantities to be different.

On the other hand, holding stock is discouraged because:

- Holding stock implies that more working capital will be needed.
- Having and operating a store requires renting or purchasing premises, paying for insurance, staff, energy, equipment, racking, and so forth. Usually the more stock that is held, the higher the costs.
- Stored stock may suffer obsolescence, deterioration, damage, and pilferage.
- Having stock can mask problems that would otherwise be identified and possibly eliminated.

The last reason and other disadvantages, such as the possibility of stock being left over when demand falls, encourage the use of just-in-time (JIT) methods for items with certain demand characteristics. JIT aims to receive items just before they are needed rather than having the items available just in case. JIT is usually considered to be a demand pull system that emphasizes that inventory is waste and suggests that stock can mask problems such as long setup times, poor quality, unreliable equipment, and unreliable delivery.

If the inventory system operates in very uncertain conditions, the system needs to be designed accordingly. If the uncertainty arises because of poor quality supplies and uncertain timing of deliveries, the problems can be overcome progressively because they are repetitive organizational problems. This then changes the emphasis from holding stock to protect against the uncertainty to eliminating the problems. If that can be done, the need for stock is reduced and the result can become effective JIT; that is, a steady flow of items through the supply and distribution chain that arrive at the required location at the required time without accumulating items at intermediate locations. JIT descriptions frequently suggest the need to develop a network of suppliers that are located geographically close to the user so that deliveries can be made rapidly. Similarly, reducing long setup times allows items to be made economically in smaller quantities and so speeds up the organization's response. The quicker the response and the better the quality of the parts, the easier it becomes to base manufacturing quantities on real demand, and the closer the system comes to truly being a demand pull system. Nevertheless, JIT will be unsatisfactory in many situations, particularly where there is a lot of uncertainty, for example, with long lead-time items with irregular demand and where demand responds to visibility, as in retailing. Stocks will be necessary then, even if JIT is used in other parts of the logistics chain.

Uncertainty can arise in a particularly acute form in disaster situations, such as those posed by tsunamis and earthquakes; military conflict zones, particularly civil wars with associated refugee problems; new medical conditions, such as epidemics; and so forth. All of these are inherently difficult to anticipate and stocks are needed so as to be able to quickly respond. However, because of so much uncertainty, JIT as described is inappropriate. Indeed, planning for disasters may require the use of entirely new inventory paradigms. These will not be discussed further here.

The discussion now considers the more common stock situations and examines the different stages in the logistics chain. These stages identify where stocks may exist. The discussion then considers some simple traditional inventory analysis. Environmental analysis is deferred until Sections 3.3 and 3.4.

3.2.2 THE LOGISTICS CHAIN

The logistics chain for stock items is taken to include material supply, manufacture, product storage, and product dispatch for customer use. Thus, within a logistics system the main stages for manufactured items include: mining the minerals (e.g., coal and iron ore); converting the mined material into raw material (e.g., steel); making piece parts; assembling these parts into finished products (this may require constructing several levels of intermediate subassemblies); testing the products; packaging the products; distributing the items to warehouses and other storage areas, wholesalers, and retailers; selling the goods; and sending them to the eventual users. From the environmental point of view, particular interests focus on how the product is produced; how it is used; and, after use, what actions are taken to ensure that the items are appropriately scrapped, reused, or recycled. Product design, process design, and manufacturing system design have an important influence on most of these steps. Additionally, there will be transportation between the different locations within the chain, and there may be storage at the receiving and dispatch areas of any location. Emphasis on the total system is important from an environmental point of view.

In practical terms, the use of specific suppliers for some items may be, at least in the short term, almost inevitable because of material and skill availability. For example, the location of mines or specific manufacturing skills or overriding economic factors may mean that it is much cheaper and more effective to transport goods large distances than to produce them locally. An extreme example of this is the growth of crystals in space where quality conditions may justify additional cost. More prosaically, there have been some circumstances in recent years where it has been cheaper to buy completed products from China than to purchase the material to make the items elsewhere.

3.2.3 CONVENTIONAL INVENTORY ANALYSIS THAT DOES
NOT CONSIDER ENVIRONMENTAL FACTORS

Some simple inventory rules that do not consider environmental factors will now be discussed. The simplest inventory control models consider inventory flows into and out of a single stock location. The inventory control rules respond to the demands for the item by eventually creating a replenishment order. However, if the replenishment or manufacturing lead time is longer than the available demand knowledge, it will become necessary to forecast the demand and to start manufacturing to meet the anticipated forecast demand.

In principle, stocks could be held before and after each stage and operation within the manufacturing cycle. Additionally, process stock may accumulate at each operation and some of this may be transferred onto the next stage or operation. Thus stocks may be in transit between each operation. The same principle applies to the complete logistics chain for a product that could include many raw materials suppliers, multilevels of assembly, and products that are dispatched to many customers or warehouses.

Probably the simplest inventory control model is the continuous review inventory control system; that is, the stock is reviewed after every transaction. If it is assumed that the demand is uniform, that there are no stockouts, and that the lead time (L) is fixed with delivery occurring just as the stock is about to run out, then it is clear that the system repeats itself and there are no shortages. In this case it is straightforward to derive the reorder level (ROL), s, and the reorder quantity (ROQ), q, that minimize the total cost of the operation. For example, if the ordering cost and surplus cost are known, then the optimal value for the reorder quantity, q, may be calculated using the Wilson lot size formula:

$$q = \sqrt{\frac{2Ar}{h}} \qquad (3.1)$$

where A is the order cost, r is the demand rate, and h is the holding cost per unit per unit of time. The value of the reorder level is $s = Lr$, where L is the lead time.

A slightly more realistic situation occurs when the demand exhibits variability about a constant mean. The reorder levels and reorder quantities can be derived more or less as before, but now, because of the variability, each cycle of the system will be different and a safety stock is needed to provide protection against the random variation of the demand. In this case the classical analysis balances the cost of satisfying a proportion of the demands (by holding safety stock) against the cost of shortages. The performance is a statistical measure that on average the system will aim to satisfy, say, 95 percent of individual demands.

Inventory problems are somewhat greater when there is more than one stage in the logistics chain, which of course is the usual situation. Interestingly, rules that appear sensible for a single stock location may amplify demand when applied to a process with two or more stages. Indeed, as a system becomes more complex, the dynamic system performance may start to exhibit unexpected characteristics. For example, in a simple two-stage process where each stage uses the simple EOQ model, the batch size representing the demand gets progressively amplified.

It is common to combine inventory control rules with demand forecasting rules if it is considered likely that the mean demand will change with time. The rules for forecasting demand, for planning and controlling manufacture, and for planning and controlling inventory depend on the product and the nature of the market demand. If the demand changes rapidly, there is a need to identify trends, seasonality, and so forth, and build these into the forecasting system. A frequently used demand forecasting system that adjusts well to (not too rapid) demand changes with some randomness is the exponential smoothing system. Whatever rules are chosen, the rules, together with the lead times, respond differently to the demand that the system is trying to satisfy. In other words the dynamic performance of systems is affected by the use of different forecasting systems and different inventory control rules that determine when and how much to order. The response is also affected by data such as different lead times and different demands.

Many models have been analyzed (e.g., see Chikán [1990]). These models correspond to the consideration of many different factors and different statistical distributions of demand. Each model is of course a simplification of reality and considers only a limited number of the possible variables that face an inventory control system designer. A fuller description of inventory control and its many rules will be found in standard texts, such as Silver et al. (1998).

3.2.4 DESIGNING AN INVENTORY CONTROL SYSTEM

An important problem is to decide what inventory control system should be chosen so as to respond appropriately when the system is subject to various types of demand.

In the context of this chapter we want to choose inventory control rules that are workable, have satisfactory dynamic performance, and also give good environmental performance. We want to provide good and consistent service and we want to ensure that holding stock is a good investment. Among other things, we need to determine the inventory parameters such as the reorder levels and reorder quantities, and the performance of the inventory control system over time. If we wish to make and maintain complex products such as automobiles and aircraft and their subsystems, it is essential to use systematic inventory planning. This is often achieved within a context that is based on material requirements planning (MRP) or enterprise resource planning (ERP).

Fluctuations in stock levels and in production levels can lead to shortages, rush orders, or surplus stock, each of which has economic and environmental implications that need to be better understood.

Among the key questions that need to be answered about the inventory planning system include:

- Are products to be made to order or for stock? The answer to this is largely determined by the delivery lead time required by the market and the manufacturing lead time of the suppliers. If the delivery lead time required by the market is less than the manufacturing lead time, then it will be necessary to hold stock. For example, if the manufacturing lead time is twelve weeks and the required customer lead time is four weeks, there is a need to start manufacturing in anticipation of future sales, and in this case to have eight weeks of stock in the pipeline. This obviously needs to be based on a forecast of customer demand and it implies that there will be stock held at various stages in the production sequence. Stocks held at different stages of the logistic chain allow appropriate service to be provided. If the forecast of what is a week's stock is incorrect or demand subsequently changes, there will be either a shortage or a surplus. For example, if the demand increases above the forecast then the consequence could be a decline in the planned service, whereas if the demand falls that could leave surplus stock.

- What rules should be used for controlling the stock? These will include any rules for forecasting, rules to determine when and how much to order, and rules for controlling production.
- Where should the stores and depots be located?
- Where should the items be made or bought?

To design a logistics planning and control system that has the desired least cost performance is difficult, but once the system is installed, observation and measurement can identify specific instances of the problems that could arise. However, to be more systematic, mathematical and simulation models of the possible inventory systems could be constructed and used to evaluate the performance of a proposed system before the system is installed. Specifically, the response of the proposed system could be tested by putting both typical and unusual demands into the system model. Such tests could include checking the response to increased or reduced demand, checking the effect of occasional large orders, checking the effect of deliveries that arrive late or are of poor quality, and so on. Examining such instances encourages users to plan contingency responses to potential situations. This analysis should be extensive and may be quite time consuming for A and B items that would be chosen by using the well-known Pareto analysis (ABC analysis) i.e. high and medium usage-value.

3.2.5 SUMMARY

Section 3.2.4 indicated the type of information (essentially the demand, the stock level, the costs, and the time that it takes to transport and manufacture items) and decisions that are required to define a traditional inventory control system. The methods could be chosen based on rules of thumb, mathematical analysis, and simulation possibly combined with local expertise that takes account of local conditions and the expertise and experience of the current operators. In other words, there are many advantages in combining mathematical analysis, simulation analysis, and practical studies together in a mutually supportive manner when examining the performance of traditional inventory systems.

As we shall see later, because of the need for agreed performance measures, and because the environmental timescale is unlikely to coincide with the operational activities of the inventory system, it is even more difficult to measure the environmental performance of an inventory system. Therefore, when designing the more complex environmental inventory systems, it is even more desirable to use a combination of modeling techniques to investigate the outcomes.

3.3 AN ENVIRONMENTAL CONTEXT FOR INVENTORY

3.3.1 INVENTORY SYSTEMS FACTORS WITH ENVIRONMENTAL IMPLICATIONS

Section 3.2 presented a broad statement of the inventory control problem and outlined some simple traditional inventory control methods. The discussion now considers individual logistic factors that potentially have positive or negative environmental implications and how they may interact with inventory decisions. The aim is to see whether this knowledge can help us design inventory systems that take account of environmental implications. Although it is environmentally very important to consider how a product (e.g., a car) is used, this is not considered here because it is not under the control of the inventory planner.

The discussion considers the operation of the logistics system and the steps involved in the lifecycle of the product including product production. Some of the factors that interrelate with inventory systems and that have environmental implications are:

- Product design
- Process design
- Product manufacture
- Transport
- Alternative products
- Obsolescence
- Waste
- Packaging
- System induced effects
- Inventory storage

Bonney et al. (2000) discussed the product introduction process, which was taken to include product design and process design, together with the manufacturing systems design that is needed to control the production. The manufacturing systems design included plant layout and design, and facility design and evaluation. Bonney et al. suggested that an extended concurrent engineering (sometimes called a concurrent enterprise) viewpoint could help to coordinate these different stages of the product life cycle.

The listed inventory systems factors, including the product introduction steps in the product lifecycle, are now considered. After considering these individual factors, the chapter proposes metrics to measure them and it then examines them to see whether they could form the basis for investigating the environmental effects of inventory and encouraging greener activities.

3.3.2   THE INVENTORY SYSTEM FACTORS

3.3.2.1   Product Design

Product design is probably the most important step in determining how resources will be used. Product design aims to create a product that will satisfy the market requirements in terms of function, quality, cost, and time. Product design—aided by appropriate tools such as value analysis, value engineering, standardization, and so forth—determines the materials used and the tolerances; in other words, how the product will be made and the product cost. Specifically, value analysis and value engineering may provide a helpful basis for simplifying the product, minimizing the number of parts in a product, and minimizing the environmental implications. It is advantageous to build all of the planned requirements into the product at the design stage rather than attempting to modify designs later. How design relates to manufacture is discussed next.

3.3.2.2   Process Design

Process design involves the use of standard parts (whenever possible), the choice of materials, the choice of machines and the processes used, jigs and fixtures, labor requirements, workplace design, and so forth. Designing for manufacture can be used to minimize the environmental impact of the production processes that are needed.

3.3.2.3   Product Manufacture

Product production is particularly important from the environmental point of view. A manufacturing organization aims to convert materials into products efficiently and then to distribute the products to customers and storage locations appropriately. The efficiency of the product production is largely determined by:

- The product design
- The process design
- The design of the manufacturing system that makes the product. This manufacturing system includes planning the material flow, choosing the planning and control methods including production and inventory control, quality control and financial control, the availability of appropriate stocks and resources to meet the demand, the organization used to achieve the conversion of materials into products, and the methods of operational control.

For products that are to be made for stock, the production/order requirements for products, assemblies, parts, and material will be determined by using demand forecasts, the current levels of stock, the inventory control rules, and the production planning rules. For production that is made to order, inventory may rise as a result of changes to production quantities. In both situations stocks of materials and parts may be held for the reasons mentioned earlier and these may also be used to help to reduce delivery lead times. Environmental effects result from: the planning process, changes in demand during manufacture particularly at the start and at the end of a product's life, the efficiency of the production processes, the scrap and waste that are created, and the pollution that occurs during production. Changes in the physical quality and times achieved may lead to over- or underachievement of programs, quality targets, and so forth, and these may have environmental implications.

3.3.2.4 Relationship between Inventory and Transport

In the early years of the Industrial Revolution, goods were usually produced close to the source of the essential resources such as iron and coal. The finished goods were then transported to the markets. However, spinning and weaving in the textile industry in the United Kingdom were greatly influenced by climatic conditions; for example, damp conditions in Lancashire for spinning and drier conditions in Yorkshire for weaving. Later, materials such as coal and iron ore were brought to where the skills and the infrastructure of roads, canals, railways, and distribution systems were located.

Nowadays, the location chosen for production is more likely to be dependent on the perceived total cost of production, including labor costs, the availability or potential availability of skills and the education and training level of the workforce, and the available support infrastructure, which includes transport and communication systems. Also important is the willingness of the community, government, and people of the area to accept the production facilities and associated infrastructure, including possibly detrimental environmental consequences. Many developing countries will use potential employment opportunities as an excuse to develop their national infrastructure of roads, railways, and ports (sea and air).

Reasons for transporting goods include:

- Production efficiency—It may be more efficient to manufacture large quantities in one location and then transport the items to where they are required rather than transport the materials to near where they are required and then manufacture in smaller quantities locally.
- Skill—It may not be possible to obtain the required skills locally.
- Environmental—It may be better to refine things before sending rather than after because items will then be lighter and the transport volume and cost may be less. It may also be easier to reuse any waste such as offcuts and other scrap metal by reprocessing it at the source.
- Cost—Despite goods traveling greater distances, the sum of the travel and manufacturing costs may be lower because of lower labor costs or lower material costs. However, sometimes the apparent lower costs may be based on using accounting methods that do not take into account the full cost of travel or the effect on local communities. This chapter specifically argues that the travel costs that are used should include environmental costs. Currently, for reasons already outlined, it is highly unlikely that the company concerned will take account of the environmental effects of the travel.
- Climatic—Sometimes processing may be easier in a warmer than in a colder climate and vice versa. Specifically, the climate also may affect the energy costs associated with production and storage.

- Quality—It may be possible to have fewer spinning faults in a damp climate. Another example is that better quality fruit may be available for canning or freezing if the processing factory is close to the growing areas.

Holding stock at a specific place implies that there will be travel involved in taking items to and bringing items from that location. Different locations will have different associated financial and environmental costs, and, therefore, there is a need to choose appropriate stock locations and suppliers to minimize the total environmental (travel and infrastructure) costs. With this as an objective, there may be an advantage to restrict sales from a particular source to specifically defined geographical areas. This may become a rigid company policy or one that is encouraged by appropriate pricing so that environmental and organizational costs are minimized. For example, reducing travel would be encouraged by companies having a green tariff that charges reduced rates for local deliveries by being part of a regular schedule, and charges premium rates for longer distances, urgent deliveries, and environmentally damaging travel.

With retailing, there will be less customer travel if the shop is close to the community that it serves. On the other hand, large out-of-town supermarkets necessarily generate travel, and current supermarket locations are frequently based on where large sites are available, the cost of land/rent, and so forth. When shops are located in city centers, much inconvenience arises from congestion, parking, and so on. With distribution centers there is the need to choose the location and also decide whether to have multiple distribution centers or a single center. Internet and mail-order retailing and the method of delivery that is used also affect the environmental costs of collecting and delivery.

Analogously, if local production is encouraged, the amount of travel that is required is reduced. With perishable items such as food, there may be additional advantages such as improved quality. Marketing now likes to highlight when something has been locally sourced as this plays on local loyalties and implied quality. On the other hand, reduced dependence on local sources and local manufacture by using sources that are farther away implies greater transportation of goods and, in general, longer delivery times.

Specific situations to be avoided include when similar products are sent in similar quantities between different areas or countries, for example, bottled water between different countries. Except in times of humanitarian disaster, bottled water almost universally is considered environmentally bad because of its production methods, its packaging, and its transport. However, a major deterrent to reducing this trade is the very clever marketing that means that many people now perceive bottled water to be of better quality than tap water. Also, selling bottled water is highly profitable. This raises the question whether rules designed to encourage environmentally good actions would be considered to be an infringement of individual freedom or a constraint of trade placed on restaurateurs.

3.3.2.5 Alternative Products

Alternative products are available in many consumer and industrial situations. Some of these items may have a much lower environmental cost than others but are of similar functionality. Their use should be encouraged and information about such situations should be available so that designers and users can make informed judgments. However, with food products, in addition to the local sourcing issue discussed earlier, there are also many products whose withdrawal could have positive health benefits. It can be seen that if this information was formalized into environmental rules, then this would overspill into actions that would affect commercial decisions and personal choices and achieve little except resistance to the idea of external involvement. Again there is the potential to affect freedom of choice, and care needs to be taken in such situations.

3.3.2.6 Relationship between Inventory and Obsolescence

Obsolescence is a stage in the product life cycle when, for whatever reason, the end of the product life is about to arrive. The time when this occurs depends on the perishability of the product, fashion, the dynamics of the market, external influences such as technological changes (design methods, rapid prototyping, etc.) and material changes (such as the availability of different colors, textures, and strengths of new materials), health and safety requirements, environmental influences creating change, and the economic trade-off between reusing or throwing away items.

A difficult problem is to decide whether an item should be scrapped rather than refurbished, particularly when the performance of a refurbished machine may be poorer in terms of productivity, environmental effects, and so forth. A new machine could have lower energy consumption, more precise operations, and produce less waste.

Obsolescent stock is the set of items that is still usable but exists unused at the stockists at the end of the product life cycle. The disposal of this remaining stock may itself have an environmental cost. The stock levels at the time when a product model changes need to be planned and controlled so that they are appropriate to the costs and risks that their shortage or surplus incur. It may be also possible to use some excess stock for spare parts, but ideally spare parts provisioning should be planned as part of the transition procedures. Generally, the sales potential of obsolescent stock declines rapidly as a new model becomes accepted, particularly if there has been a major technological improvement or fashion change. Influences on obsolescence include:

- In general, the longer that stocks are held, the more difficult will it become to sell the items.
- Imbalance in the stocks that are held; for example, if there are more right-handed than left-handed items available and the items are normally sold in matched pairs, it is likely that stocks will be leftover and some items will need to be scrapped.
- Technological improvements
- Fashion changes

Potential obsolescence is a major reason why retailers have sales, particularly of seasonal or fashion items such as clothing; a reason why car producers sometimes produce special edition models that include more extras than normal; and so forth. In situations where items cannot be sold, often because of a complete change in the fashion or technology, then the leftover products may need to be scrapped and become part of waste (see the next section).

3.3.2.7   Relationship between Inventory and Waste

Waste is somewhat more general than obsolescence and is affected by the perishability of the product. Waste can arise from incorrect ordering, items that have been damaged in store, or items that have corroded or perished because of poor storage or packaging. Waste also occurs when, to a specific user, the effort to recycle or use for another purpose is perceived to be uneconomic. Examples of this arise with builders whose wood scraps or small quantities of items are leftover from the immediate job in hand. Much food is wasted because of being left unused for too long. At the end of a day, cooked food in restaurants is often scrapped for health and safety reasons.

In many circumstances waste can be reduced by creating appropriate storage conditions, for example, by controlling temperature and humidity. This might include refrigerating or freezing items to extend their life. Such improved storage can allow items such as fruit to be available year round. Influences that increase waste include:

- Items becoming damaged
- Deterioration of the items
- Obsolescence of the stock held
- Poor organization either by the supplier or the potential user

3.3.2.8   Relationship between Inventory and Packaging

Packaging is often required to protect or handle a product and so it is linked to product design. Packaging is commonly required to reduce damage in transit or provide security against theft. Eventually, this packaging usually becomes waste. For this reason, there have been attempts to make packaging recyclable and to place the responsibility for the cost of its disposal on to the producer of the goods who chooses to use the packaging; the additional cost should encourage designers to make more effort with their packaging design. Another example is the use of plastic bags for convenience of carrying or for marketing. A recent consumer-led initiative to reduce the use of plastic bags (some UK supermarkets probably encouraged the initiative for environmental and cost reasons) provided incentives for customers to use their own bags or to recycle the old. Another approach that has been adopted by some organizations is to charge for the use of bags.

Ways to reduce packaging requirements include:

- Design products to minimize packaging
- Build some of the packaging function into the product
- Design the packaging so as to minimize environmental impact, for example, by using recyclable or biodegradable packaging

- Use multifunction packaging so that it becomes part of the labeling and product selling function, for example, on the cans of drinks
- Design packaging that can be reused or used for other purposes

3.3.2.9 Relationship between Inventory Systems and System-Induced Effects

It was commented earlier in the chapter that system-induced fluctuations in stock levels can result from applying the inventory control rules. This section outlines how this occurs.

There is a great deal of knowledge about how to analyze the performance of control systems, particularly those based on electrical, electronic, and mechanical components. Without this knowledge, the performance of automobiles, aircraft, and much military equipment (including radar systems and missiles) would be poorer. The basic principle of a control system (also known as a servomechanism or as an error-actuated feedback device) is that the difference between the desired and actual performance of the system at any time is used as a signal to change appropriately whatever is being controlled. Simple examples of control systems are the temperature of a room being controlled by a thermostat, the distance from a target or the angle off from the target being used to control the path and response of a missile, or the depression of a spring that may, among other things, help to stabilize the suspension of a car. The behavior of a physical control system is frequently represented by means of differential equations. Analysis of the system representation can show the dynamic performance of the system including the transients, overshoots, and steady-state solutions. Many management control systems use the same principle of having a performance target (e.g., the desired quality in a quality control system, a standard cost in a budgetary control system, the planned inventory in an inventory control system, or the required production quantity in a production control system) against which achievements are monitored and used to adjust actions.

The major differences between management control systems and hardware systems are that management control systems may be less predictable because some of the system components are people who may have motivations that are not necessarily in complete accord with the system designers; indeed, these people are likely to display intelligence and variability and they may be motivated by information provided by the system. Another difference is that the controllers of management systems are usually progressively trying to improve the system performance. However, despite this, control system ideas have been used by many investigators to analyze and design management systems so that they will have the desired performance. Tools that have been used include Laplace transform methods, z-transform methods, modern control theory, and, of course, simulation.

Work of interest in this area includes Forrester's (1961) work on what he calls industrial dynamics. Meadows et al. (2004) examined what would happen to the world's resources if the then-current policies continued. Work relevant to production and inventory control include Laplace and z-transform methods. Work that uses control theoretic models applied to the logistics area includes Towill's (1982) continuous models, particularly of inventory and logistics systems; Grubbström and Molinder's

(1996) analysis of matrix representations of material requirements planning (MRP) models using Laplace transform methods; and Popplewell and Bonney's (1987) analysis of production, inventory, and forecasting systems using z-transform discrete models to represent systems that are planned on a periodic basis. More recently Dejonckheere et al. (2003) used the approach to study the bullwhip effect.

Many of these studies show that different rules used for controlling inventory and production levels in a logistics chain can lead to fluctuation in the material held and in the production levels, and that this can lead to potential waste. However, good design can reduce the amount of variability and create stable systems that have "appropriate" performance.

3.3.2.10   Inventory and Its Storage

This is largely an architectural problem. There is a need to produce a functional building that is well insulated, easily accessible, matches the road infrastructure so that any extra traffic generated by operators and deliveries is appropriate for the location, and so forth. It is desirable that the building is designed to minimize the amount of power and resource involved in constructing and operating the stores, and the size is influenced by the levels of stock required. Delivery frequency affects the traffic volume, and, as discussed separately, the relative locations of the different parts of the logistics chain can change the amount of transportation required.

3.3.3   Major Risks Facing the Inventory Planning System

Major risks associated with inventory items arise in the following situations:

- At start-up, because lack of knowledge makes it difficult to forecast customer demand. As a result there can be too much or too little stock and too much or insufficient capacity to manufacture the items.
- When major changes are needed to the product, for example, for marketing, technological, or safety reasons. The effect could be that an organization is left with a lot of obsolescent stock.
- The dynamic effects of the inventory system can create surpluses and shortages. Shortages may require taking rush remedial action, for example, a priority delivery that may need an additional journey that uses extra energy.
- Close down procedures may leave too much or too little stock, again with the danger of leaving obsolescent stock or needing rush remedial actions.
- Major changes in the demand characteristics such as trends or steps in demand.
- Data uncertainty, particularly with costs, environmental metrics, and lead-time data. Manufacture and material-supply lead times are not necessarily constant and this may lead to surpluses or shortage of stock and require the kind of remedial actions already mentioned.

Other risks (and opportunities) arise from noninventory events. For example, natural occurrences such as floods and adverse weather conditions could affect transport availability. External technological changes, including computing and communication changes and improved quality of data, might improve the functioning of the operational inventory control system. The inventory system will need to respond to any changes.

3.3.4 Assessing the Environmental Effects of Inventory

To understand the relationship between inventory and the environment better, there is a need to be able to measure the environmental performance of the total system. In other words, system performance should be expressed in terms of an environmental cost, which includes the effects of producing, storing, transporting, using, and scrapping the product considered over the complete product life cycle and the whole logistics chain. This could be described as life-cycle environmental costing.

Much work is needed to develop appropriate metrics. To date most organizations have made very little assessment of the environmental impact of holding inventory because, other than broad measures like the carbon footprint, there is little agreement about environmental performance metrics. On the other hand, if metrics (hereafter called the environmental cost) that estimate the environmental impact could be produced and agreed upon, then almost certainly models similar to the classic inventory models could be developed and used. Indeed, whenever there are trade-offs and appropriate information can be obtained, it should be possible to rework the standard inventory analyses to minimize the metric representing the environmental costs. To measure system performance, simulation models could consider the effects, say, of different levels of energy use or of transport use.

Some suggested approaches related to potential problems are listed in a sequence that roughly relates to progressively improved environmental understanding:

- Problem avoidance and problem measurement
- Comparative analysis
- Consequential analysis
- Adverse effects analysis
- Modeling

Each is briefly discussed next.

3.3.4.1 Problem Avoidance and Problem Measurement

Problem avoidance and problem measurement assumes that, other things being equal, it is sensible to reduce activities that potentially have adverse environmental effects. Thus, specific aims of the proposed changes could be to:

- Reduce the use of resources, for example, of minerals and energy.
- Reduce the production of effluent, for example, by using different processes, better filtering, better work procedures, and so forth.
- Avoid the use of toxic materials. If toxic materials are found to be present, then as soon as is practicable, use procedures to alert their presence with

possible recall. If necessary, these procedures should be international and any cover-up must be avoided.

- Choose manufacturing and stores locations that help to minimize travel.
- As far as possible, reduce the environmental implications of transporting goods from the known stores' locations (e.g., use routing and scheduling activities to reduce the distance traveled, choose appropriate modes of transportation, etc.).
- Produce products in a way that will minimize environmental penalties in production, when used, or when modified or scrapped at the end of their life. This may require planning for refurbishment and disassembly, minimizing unnecessary processing and waste of all kinds, and producing products using efficient methods that work well and produce goods of quality.
- Treat items appropriately so that they are not damaged in the stores.
- Avoid spills.
- Avoid dumping.

Many of these items could be monitored or fully recorded, and this could later become the basis for more detailed actions. For example, moving through the afore-mentioned list, one could measure the energy expended by metering, one could check the quality of effluent, and the number of journeys undertaken could be recorded, as could the fuel consumed. Some of these points are developed further later.

3.3.4.2 Comparative Analysis

The data from the previous section shows what the system is doing and whether it is improving. With comparative analysis, the aim is to take this further.

Comparative analysis compares the effects of varying the main factors for the pro-posals under consideration (such as the locations of manufacture, travel method, plan-ning methods, etc.) on the full logistics chain. Possible performance measures could include the consequences such as cost, quality, travel distances, the time that it will take to transport, uncertainties, and so forth. These performance measures would then provide a factual basis for making intuitive decisions, even though, because of differ-ent weightings, not everyone will necessarily come to the same conclusion. The more knowledge that is obtained then the closer the analysis comes to being a simulation.

3.3.4.3 Consequential Analysis

This approach uses modeling to better understand the consequences that would arise from hunches. The more detailed that this becomes, then the closer the approach moves toward formal modeling. Where factors are found to have a dramatic effect on inventory parameters, this suggests that it may be worth investigating the relation-ships more thoroughly.

3.3.4.4 Adverse Effect Analysis (Modeling)

If, in given types of situation, it is possible to make a realistic assessment of the environmental consequences of an inventory decision, then it should be possible to analyze specific instances of the problem avoidance list. Many of the initial actions could be classified as good housekeeping and are probably worth doing in their own right. Specific instances of this would be to examine production methods or to examine design methods. These would almost certainly pay for themselves even before any environmental analysis was performed. However, with environmental analysis, the effect on resource usage (indeed the effect on most of the problem avoidance measures) would be a routine consequence of the analysis. The consequence of each analysis provides data relating to another situation equivalent to another case study. Gradually, these would lead to a progressive increase in understanding. This in turn would help to formulate the procedures to be adopted. This data could eventually become the basis for modeling.

3.3.4.5 Modeling

It might have been inferred from the thrust of the argument that modeling may become more possible as knowledge improves. The author thinks that this is a desirable objective that would improve understanding progressively. Modeling is a practical possibility in many situations. Based on sensitivity analysis, it may be possible to identify some factors as second-order effects and then to eliminate these from further immediate study. The effect of combinations of factors and a variety of measures can be modeled for subsequent analysis and discussion. Hopefully, this would lead to obtaining agreed upon composite measures. Modeling, once based on the use of these agreed upon composite measures, can be used again.

Another use of modeling is to evaluate conjectures in sufficient detail to see whether the consequences of the conjecture are substantial. An example of this is whether a measure such as an entropy cost could be used to partially represent the cost of maintaining control of systems, for example, to avoid system quality problems (Jaber et al., 2008). The aim is initially to check whether some representation, if reasonable, will have important effects, and then to provide guidance on what is needed for further investigation.

It is assumed that modeling can eventually be used to derive a realistic assessment of environmental effects in many types of situations. This would allow specific models to be created that may be used to minimize the error in the estimate of that composite measure, which was earlier called the environmental cost.

3.3.5 Metrics

Most of the actions associated with the aforementioned aspirations may be recorded and the simple consequences monitored, for example, the use of resources, the amount of effluent, the use of toxic materials, the travel consequences of changing the locations of manufacturing and stores, and so on.

The metrics mentioned so far have been somewhat simplistic. They provide a factual basis for auditing and for action, but they do not provide a sufficiently good environmental measure to rank activities or to provide an environmental cost or a decision algorithm. Indeed, it is probably not realistic to rank different actions expressed in terms of an agreed upon single environmental metric. However, the measures can be the basis for interpreting situations and improving the quality of environmental decisions. These ideas are now developed further.

3.4 AMALGAMATING INVENTORY AND ENVIRONMENTAL CONSIDERATIONS

3.4.1 DEALING WITH INVENTORY AND ENVIRONMENTAL INTERACTIONS

Potential environmental savings are most likely to occur where inventory and environmental considerations interact. Bonney and Jaber (2008) considered the main categories of players who influence inventory use and control. The paper suggested that there are five categories of players related to inventory and the environment: international organizations, nation states, local governments, companies and other organizations and, finally, people. The categories are outlined in Section 3.4.2.

There is a need for systematic and easily usable methods of environmental inventory modeling, as this would allow us to assess environmental consequences and hence give environmental considerations high priority. Specifically, these methods need to recognize that the timescale over which environmental problems manifest themselves is often much greater than the normal timescale of management decisions. (Of course timescales could also be shorter, but in that case the environmental problems would usually be identified and eliminated during product testing.) Equally important are to be able to make good commercial decisions that do not have adverse environmental consequences, and to make good environmental decisions despite possible short-term adverse commercial consequences. In short, decisions need to be profitable but we also need to leave a world that is fit for future generations. When there is a trade-off between commercially and environmentally based decisions, then environmentally good activities need to be encouraged. To achieve this in practice will probably require a combination of education, commitment, incentives, and penalties.

If it is possible to represent environmental considerations approximately, simply by changing the costs, then one would expect the inventory analysis and consequent derived rules to have basically the same structure as traditional analyses. If, on the other hand, more complex formulations are needed to represent environmental implications fairly realistically, then the model structures and the ordering rules that are derived may need to completely change.

After considering metrics in greater detail, environmental inventory models are investigated. First, however, the players and their responsibilities for activities that potentially affect the environment are discussed.

3.4.2 CATEGORIES OF PLAYERS INVOLVED

Environmental effects and resource usage do not necessarily have geographic boundaries. Polluted air and water may travel many miles from their source. Shortages of resources can affect everyone with changes to resources' availability and prices. The effects could trigger disputes and, in extreme cases, could require international agreement to reconcile the differences. For these reasons it is important to identify who has responsibility for specific activities and to create frameworks within which all the parties (players) can work. The following is an attempt to structure the relationships between the players.

In their brief examination of the environmental needs for inventory systems, Bonney and Jaber (2008) considered five (hierarchical) categories of players who have an influence on the environmental implications of inventory. The five players and their roles were suggested to be:

- International organizations, which should set frameworks within which nations, by agreement, will work.
- Nation states, which should first identify and balance international needs with the needs of their own nationals and enterprises; and, second, should provide rules and legislation that set the context within which players at the lower levels of the hierarchy would operate.
- Local governments, which should interpret and implement the national rules (for example, those related to building regulations, possibly adjusted in the light of local conditions), and guide local enterprises and individuals toward better environmental solutions.
- Companies and other organizations, which should be "as green as possible" consistent with their other organizational requirements such as needing to work within the law, make a profit, and provide a good working environment for their employees.
- People, who need to act responsibly within the frameworks set by the other players, but who should also aim to influence decisions within organizations with which they have contact as employees, members of professional bodies, customers, etc.

Of the five players, the first three set the rules or frameworks within which the others work, but it is the companies and individuals that specifically make the decisions that choose and use inventory items. For these reasons this chapter focuses on providing guidance specifically at the company level.

3.4.3 PRINCIPLES UPON WHICH TO BASE RESPONSIBLE INVENTORY DECISIONS

The environmental consequences of some actions are potentially so great that there is a need for organizations to develop formal procedures to sign off new proposals for systems, workplaces, and products. The procedures need to be systematic but expeditious. They should not be used to slow developments, as there will be an ongoing need to introduce new solutions quickly as better knowledge is obtained. Ideally, the procedures should be internal to the organization but open to inspection from the time of certification. In other words, it is suggested that an inventory planning system that was designed to take account of views on all of the factors that were discussed in Section 3.3 should be approved using agreed procedures that retain some flexibility. They should be more like agreeing quality systems, health and safety systems, or auditing procedures, rather than systems such as the Swedish system for workplace design, planning procedures for local authorities, or, even more exacting and costly, drug approval in medicine.

A tentative list of principles upon which decisions could be made include:

- Identifying who is responsible for what (the players and their responsibilities are outlined in Section 3.4.2). This needs translating into who in the management function is specifically responsible for designing the system, who will be operating the system, and who is administratively responsible for checking that the system is operating effectively. Problems that arise need to be quickly rectified to the satisfaction of local operational requirements and the wider environmental requirements.
- Ensuring that decisions are economically and environmentally good. If and when there are doubts, then these should be noted. If the consequences are potentially great, further research should be carried out.
- Avoiding environmentally bad decisions.

Possible ways to check these conditions are to:

- Simulate the proposed system as indicated in the previous section.
- Pilot any radical or controversial proposals before installing.
- Monitor performance after installation.
- Regularly audit the procedures and their performance; identify any remedial actions that are required and have procedures in place to ensure that they are implemented.

3.4.4   PROCEDURES FOR CHOOSING ACTIONS

It is the responsibility of management to ensure that appropriate procedures are in place so that good actions are performed and bad actions are reduced or eliminated. This section asks whether incentives could be used to encourage environmentally good activities and penalties used to discourage environmentally bad actions.

The decision making is straightforward if a good environmental solution is also financially better for the user. The problem then becomes one of justifying the investment that is required, and then agreeing upon actions such as training staff in the new procedures, investing in the required technology, implementing change, and ensuring that the new procedures are then followed. The problem is then one of investment appraisal. It would be acceptable to use standard investment appraisal methods in this case.

The decision making is less straightforward, however, if the solution was not financially better but the environmental arguments were strong. How does one assess the investment in this situation? First, it would help if incentives were provided to encourage the good actions, and possibly sanctions to discourage the not-as-good actions; but for this to be practical the actions would need to be based on generic measures such as energy expended, distance traveled, and so on. These would need to be part of some wider agreement so that it is not to become an open invitation to bring in protectionism by the back door. Possible actions could range from setting appropriate trade tariffs, introducing subsidies that ensure that the financial or performance penalties were not too great, or introducing legislation. As we are interested in getting good results over a very wide range of items, successful outcomes would require the enthusiastic involvement of many of the population. In turn, this would require that the purpose of the activities would need to be clearly explained, and that the reasons for the proposed restraints, incentives, or penalties on the activities made clear.

3.4.5 COULD METRICS BE USED TO ENCOURAGE ENVIRONMENTALLY BETTER ACTIONS?

First, we need to be able to identify the actions that are environmentally "good" and those that are environmentally "bad." To encourage good actions and discourage bad actions will require agreement on what is good and bad and some sort of measurement to know what is happening. The history of voluntary actions is not very encouraging and so, almost certainly, most actions will need to be backed by encouragement through personal commitment/involvement or through incentives or taxation. As an example related to a slightly different area, lower taxation on energy efficient cars may encourage people to use smaller cars. Corresponding disincentives such as high tax and petrol duty on gas-guzzling cars could be used to penalize high fuel usage. This is an example of a green tax that makes it less expensive to carry out good rather than bad actions.

However, there will always be people who can afford to ignore incentives. Therefore, encouraging green actions (e.g., recycling waste) may need to be backed by appropriate laws supported by sanctions, or possibly by using community publicity to shame antisocial actions. Unfortunately, the potential unpopularity of some green taxes, such as an increased tax on fuel, might make it difficult for politicians to implement such measures. Even though indexing tax on fuel for cars had been agreed on in the United Kingdom, when that coincided with a rapidly increasing price of (crude and) refined petrol, it led in early 2008 to protests by truck drivers and motorists, and an unwillingness by politicians to face the growing opposition. In addition to such green taxes, the only other experience so far is limited to carbon trading programs that have been criticized by some for being used as schemes that allow groups to avoid environmentally appropriate behavior.

To manage most situations in a sensible and justifiable way, we need metrics to inform us about the implications of actions that have been taken, and eventually those that will be taken. Ideally the measures should relate directly to the action being measured; for example, fuel consumption is likely to be a good candidate for assessing the impact of car use and the data is available for individual car models. However, in general, great care must be used when setting up metrics and developing approaches that are intended to encourage environmentally good behavior, particularly if the measures are indirect. So far performance measurement systems seldom obtain the expected results and they often introduce the law of unforeseen consequences. Therefore, metrics will need time to be piloted, to settle down, and also to develop over time. They also need to be accompanied by appropriate educational support.

3.4.6 EXAMINING AND IMPROVING ENVIRONMENTAL PERFORMANCE

The following analyses indicate some investigations that could be undertaken to study the interrelations between inventory and the environment, and to obtain insights about the importance of the factors:

- Produce a preliminary list of possible factors and actions.
- From that list produce a possible ranking.
- Carry out ad hoc actions including good housekeeping (e.g., of product design and process design). Also use quantitative common sense (e.g., if something is wasteful, then study how to reduce it).
- Produce simple ad hoc models (e.g., travel and layout models).
- Produce simple overall models.

The procedures for producing the list and its evaluation should become more formalized as experience is gained. However, the following is a suggested investigation procedure. Each of these steps is now discussed.

A preliminary set of factors to be considered first is the inventory list produced at the beginning of Section 3.3. A second set would be obtained by systematically stepping through the stages in the product life cycle. A third would be to step systematically through the geography of the logistics chain. Additionally, any other focus that the managers, who have local knowledge, consider important or appropriate to the organization should be considered. This will produce a tentative priority listing applicable to that company.

Before a detailed investigation of possible actions starts, it is important that an organization should attempt to rank these factors. Ideally a possible ranking would be obtained by producing a Pareto assessment of their potential environmental importance. If this is not possible, local knowledge and intuition could reduce the list of possible contenders.

Good housekeeping will almost certainly pay for itself even without environmental improvement; for example, simplifying designs and the design procedures would reduce the number of stock items; reducing set up times will speed up production and increase flexibility; and improving material flow and insulating stores would all have beneficial effects.

An example of the use of algorithms to reduce travel distance when the stores' locations are known should make it possible to measure the expected effect of making certain specific actions, and subsequently to measure the actual distance traveled and the modes of transport that are and could be used.

The development of simple overall models probably needs to be deferred for more detailed examination.

3.5 IMPLICATIONS: WHERE DO WE GO FROM HERE?

3.5.1 GENERAL DISCUSSION

It is probable that the human race is on an unsustainable environmental path that is caused by its activities. The consequences are thought to include global warming and overuse of resources. Any solution will need major policy changes by governments and companies, and also changes in the behavior of the majority of the people on the planet. The longer we take to respond to the situation, the greater is the likelihood that we will move toward catastrophe. The three major environmental problems facing the world are probably: the creation of greenhouse gases, the overuse of resources (including minerals, water, land, and the living environment where discernable effects on fish, forests, and biodiversity are occurring), and pollution.

Specific problems arise because in most developing countries, but particularly China and India, environmental considerations have usually and understandably come second to economic pressures. In particular, energy has been produced largely from coal, currently a relatively dirty fuel. Against this, the "developed countries," particularly the United States, have been unwilling or unable to reduce their rather profligate use of energy. For example, the United States generates roughly 25 percent of carbon emissions although it has only 5 percent of the world's population. Countries that currently have the benefits of using energy do not want to reduce their consumption. On the other hand, countries that have not had the benefits of bountiful energy in the past now aspire to catch up. It is interesting that the model of growth throughout the world is so pervasive that it is hard to describe the situation without using words like *benefits* and *catch up*. However, finite resources do not allow the current model to continue. It would appear that not only is it going to be difficult to obtain agreement on unified actions, even the terminology will need to change. Sustainability is a minimum requirement.

To reduce unnecessary environmental damage will require a major change in the balance of activities. We need to do more good things, such as produce more renewable energy, reduce energy usage by insulating, reduce waste, produce items more efficiently by improving the efficiency of the processes, improve the efficiency of systems, use things better, design better, and reduce packaging, while avoiding bad things such as using energy unnecessarily. There is a need to save nonrenewable energy by reducing the use of nonrenewable energy sources (such as coal, oil, and gas) and reducing pollution. However, it is unlikely that doing these things will be sufficient. There is almost certainly a need for a more radical rethinking of priorities, which will provide additional factors to take account of, even for action that is primarily focused on inventory in a general sense.

Although many companies are trying to do something, unfortunately, there have been many instances of companies superficially greening their product image as a marketing tool to encourage customer purchases rather than as a commitment to adopting policies that will help the world to become a greener place. Nevertheless, the continuing focus on environmental needs appears to be transferring into real efforts in some instances. It is to be anticipated that this will continue and deepen, but whether it will be done soon enough is difficult to ascertain.

The market is progressively becoming more environmentally aware with pressure coming from many sources including:

- The United Nations, European Commission, government, and states
- Specific groups of consumers
- Pressure groups
- The media including radio, television, newspaper, and magazine reports
- Individuals

Governments are beginning to set more realistic targets, but, in general, they have been unable so far to match their achievements to their planned timescale.

Perhaps the aim should be to focus the rethinking that is currently taking place as a result of the financial crisis to find and fund ways of overcoming the even more worrisome environmental situation. Financial cooperation could be used so that countries will work together to help the planet. Specifically, the source of energy and other resources is an important part of the environmental cost. Major investment is needed. Perhaps, taxation should be levied on carbon-based energy sources, with these funds used specifically to invest in renewable energy production methods.

However, much of this is rather pious and wide ranging compared with the task that was set for this chapter; namely, to focus more tightly on inventory and its associated problems over the logistics chain.

3.5.2 QUESTIONS TO ASK ABOUT INVENTORY

It is important to understand the implications—the opportunities and risks that arise from different inventory actions. Among the questions that could be asked are:

- Could it be environmentally advantageous to change the logistics control system including the amount and the location of stocks and the manufacturing plants?
- How would the total system performance change if appropriate inventory control policies were used? Specifically, what would be the gains and losses obtained by using different types of systems?

Specific inventory questions are:

- How do we determine the best locations to manufacture specific goods?
- How do we determine the best place to hold stock? Can we relocate stocks so that total travel is reduced?
- Could holding appropriate stock allow the amount of transportation to be reduced (e.g., by using bulk deliveries or using more environmentally friendly but slower modes of transport such as rail or road rather than air freight)?
- Do reduced stock levels increase the need for urgent deliveries?
- Can one obtain the environmental and cost advantages from bulk deliveries by delivering a range of items together while retaining low levels of stock?
- What are the environmental consequences of different distribution systems?
- What are the effects of home delivery systems?
- What are the effects of Internet shopping?
- How do we prioritize the suggested actions when answers to these and other questions are obtained?

We need to obtain answers to these and many other questions about inventory if we are going to ensure that inventory planning can play its full part in creating a sustainable environment.

3.5.3 Summary and Conclusions

How inventory planning relates to the environment has been the main theme of the chapter. The points that were made included:

There are many environmental problems and human activities are contributing to these problems. The problems include:

- the increase of greenhouse gases in the atmosphere, largely caused by the use of hydrocarbon based energy, will lead to global warming and catastrophic consequences for the world;
- our excessive use of other resources means that current actions are not sustainable;
- pollution is affecting water and air quality, bio-diversity and the quality of living in many localities.

The scale of the problems is so severe that urgent, preferably coordinated action is needed to reduce everyone's contribution. However, even working to the most optimistic time scale for implementing changes, the expected environmental implications are very disturbing. They include climate change, temperature rises, water shortages, and flooding that will adversely affect everyone. It is essential that actions are taken as soon as possible by as many players as possible using every means possible.

Actions can be of the following kinds: use less energy and other resources, produce less pollution, increase the availability of renewable energy sources, perform activities more efficiently, produce better designs, insulate better, recycle more, produce less waste, etc. Also, in general, improve as many as possible of the current activities that have an adverse effect. Several ways were suggested earlier in the chapter by which better inventory planning could help the environment. Section 3.3.2 presented a list of inventory factors which by proper consideration potentially could help (or hinder) environmental performance and Section 3.5.2 presented a list of questions that need answering if one is to produce an optimal set of actions. Inventory (considered in a generalized sense) actions can make a contribution to improving the environment.

Unfortunately, appropriate environmental cost data are seldom available or easy to obtain. Therefore, at the present time it may not be possible to determine optimal solutions. Despite that, many positive actions can be undertaken and the results can be closely monitored. The aim is to do something positive while at the same time gathering data and improving our knowledge of the effectiveness of the actions. We need to learn; we need to learn quickly and then we need to use that knowledge as soon as possible to improve our performance, at least nationally but preferably worldwide.

Section 3.4.2 suggested that five groups of players had an interest in inventory decisions, namely: international organizations, nation states, local governments, companies and other organizations, and people. It was suggested that companies and individuals within these companies were in charge of the main inventory decisions and that these decisions need to be made within the context provided by the first three players.

Every means possible should be used to encourage desirable actions in the time scale involved. Comments were made about incentives, taxes and legislation. In particular, financial incentives should be used to encourage environmentally positive actions and taxes used to discourage less good activities. In addition, an extensive information campaign will be needed to encourage appropriate actions. The information will need to be of two kinds; first there is a need for general information so that everyone is aware of the problems and of the need to find solutions and secondly, there is a need for specific information about ways that environmental performance can be improved. This information will not appear spontaneously; persons will need to be responsible for the provision and other persons for the use of such information. Information provision could arise from industry wide advisory groups perhaps supported nationally. Implementation will need to be guided by people within the companies with appropriate authority and enthusiasm to ensure that action is taken. Legislation may be needed to encourage less enthusiastic implementers to take the necessary actions.

In the short run there are going to be some situations where what is the most desirable action is inconclusive because of a lack of clear facts within companies. However, all is not lost. Before we get precise answers, we can make do with rough and ready solutions but, additionally and in parallel, procedures will be needed that will steadily improve the measures and the company data that will be available upon which to make decisions in the future. We need to measure the results. In short, the suggestion is that the companies should do the obvious, find out more and then refine the data and the solutions that arise from the chosen actions.

Even with the use of analytic methods and even with attempts to assess environmental costs, the results will be strongly influenced by the numbers that are selected to represent the costs. Obtaining the figures raises two issues. As was discussed in Section 3.3 and Section 3.4, the first is conceptual and requires deciding how exactly one is going to represent the environmental cost. The second is a practical one that in addition to obtaining the basic data, the data also needs to be interpreted to ensure that like is compared with like particularly if competitors are ignoring long term environmental effects to obtain a cost advantage. Inventory decisions and all other decisions will be very difficult if the environmental costs are not known. To take account of environmental factors, it is likely that further technical developments in modeling are going to be needed and if these investigations are to be taken seriously then the problems to be solved need to be far more focused. Different kinds of inventory will require different priorities and investigations may need to be related to particular industrial sectors. Now is the time to move away from generalities to the detailed hard work that will be required to determine the needs for many companies backed by appropriate incentives and legislative framework.

The chapter has briefly outlined some of the points that need considering if inventory planning is going to play its part in moving towards a sustainable planet. There is much to learn and the lessons need to be applied quickly. An overriding principle is that progress will have been made if things are environmentally better after decisions are implemented than they were previously. However, because of assessment problems is should be recognized that there will always remain difficult decisions even when managers of projects are really trying to do their environmental best. This is not an excuse to allow the less conscientious decision makers to wriggle out of their responsibilities, it needs to be recognized that decisions will need to be tempered by humanity, economics, and the possibility of a sustainable environment as well as satisfying non-commercial human needs of community, open spaces, etc. A range of potential problems is likely to arise but they must not be allowed to inhibit actions. Among them are that analysis may point to solutions that are apparently in conflict with competition or an open doors policy and any enforcement will need to be based on comparing like with like. Another difficulty is that the time scale over which environmental effects may manifest themselves is different from the planning horizon of normal economic business decisions. If so, a business decision could, knowingly or unknowingly, select a profitable but less good environmental decision

Unfortunately, as was indicated earlier, current decision making has been determined using non-environmental arguments. Furthermore, in many cases artificial costs, taxes, import duties and subsidies have further distorted current decision making. For example, there is little consistency in taxation levels between countries or between different modes of transport e.g. travel by air does not pay the same fuel tax as car travel. In general, taxes are used by countries primarily as revenue earners although sometimes there may be an attempt to influence behavior by putting higher taxes on less favored activities such as alcohol and tobacco. However, taxes also influence behavior and so they should be part of the armory used to obtain better environmental behavior. Possibly, when desirable changes are identified associated legislation will be needed so as to bring costs in line initially with expenditure and then, eventually, with expenditure plus associated environmental costs. Complexity is a danger.

Finally, inventory even defined as broadly as in this chapter, is only one small part of a company's operations. If the world is serious about protecting its environment, a similar approach will need to be applied by each organization to each functional area including construction, estates, office work, sales teams, etc. Nothing less than a rethink with different priorities is likely to match the scale of the problems facing us. Those that do it well will contribute to the well being of the world and will also have started to face the problems that are likely to grow progressively in importance.

REFERENCES

Bonney, M., M. Head, S. Ratchev, and I. Moualek. 2000. A manufacturing system design framework for computer aided industrial engineering. *Int J Prod Res* 38(17):4317–4327.

Bonney, M., and M. Y. Jaber. 2008. *Environmentally responsible inventory models: Nonclassical models for a non-classical era.* 15th International Symposium on Inventories, August 20–23, Budapest, Hungary.

Chikán, A., ed. 1990. *Inventory models*. Budapest, Hungary: Akademiai Kiado.

Dejonckheere, J., S. M. Disney, M. R. Lambrecht, and D. R. Towill. 2003. Measuring and avoiding the bullwhip effect: A control theoretic approach. *Eur J Oper Res* 147(3): 567–590.

Forrester, J. W. 1961. *Industrial dynamics*. Waltham, MA: Pegasus Communications.

Grubbström, R. W., and A. Molinder. 1996. Safety production plans in MRP-systems using transform methodology. *Int J Prod Econ* 46-47:297–309.

Jaber, M. Y., M. Bonney, and I. Moualek. 2008. An economic order quantity model for an imperfect production process with entropy cost. *Int J Prod Econ,* doi: 10.1016/j.ijpe.2008.08.007.

Meadows, D., J. Randers, and D. Meadows. 2004. *Limits to growth: The 30-year update.* White River Junction, VT: Chelsea Green Publishing.

Popplewell, K., and M. C. Bonney. 1987. The application of discrete linear control theory to the analysis and simulation of multi-product, multi-level production control systems. *Int J Prod Res* 25(1):45–56.

Silver, E. A., D. F. Pyke, and R. Peterson. 1998. *Inventory management and production planning and scheduling,* 3rd ed. New York: Wiley.

Towill, D. R. 1982. Dynamic analysis of an inventory and order based production control system. *Int J Prod Res* 20(6):671–687.

4 Energy and Inventories

Lucio Zavanella and Simone Zanoni
Università degli Studi di Brescia
Brescia, Italia

CONTENTS

4.1 OVERVIEW

The progressive consumption of the nonrenewable energy stocks is determining wide fluctuations in the cost of energy (primarily gas and oil), influencing both absolute values and volatility, and leading to attention on the strategic role and function that these resources play in the economy and life in general. It is well known that the debate has quickly spread to encompass other issues, such as the availability of alternative resources, including nuclear power generation; the scarcity of the "classic" sources, such as fossil fuels; and the impact on the environment (e.g., Rout et al. 2008). Similarly, there has been growing interest for a more rational and responsible consumption of energy in the industrial processes and services, this being not only a fundamental issue for sustainable production, but also a strategic leverage for competitiveness, due to the increased contribution of energy to the final cost per unit produced. Inevitably, these issues will exert, and are already exerting, an increasing and noticeable influence on the management of supplies and inventories, thus opening new horizons and opportunities for research in this field.

This chapter focuses on these issues starting with several situations, already discussed in literature, in which inventory features (such as levels and quantities) are notably influenced (and frequently determined) by the energy content of the

75

inventory itself, and by the cost of the energy source. This is the case with dams (water reservoirs), gas/oil stocks and, more recently, warm/cool storage. These cases will allow the reader to perceive and understand how, why, and in which ways stocks may be linked to energy. The subsequent sections will concentrate on those other relevant aspects that cannot be neglected when the relationships between stock features and energy are considered. As a reference, the cases of the biomass supply chain (and, more generally, biofuels) and the "cold" supply of goods will be discussed. These examples will introduce the final comments, which will remind the reader that, when energy is employed, environmental aspects (e.g., emissions) are also to be taken into account. Therefore, several topics will be addressed to illustrate how, when the relationships between energy and inventory are considered, traditional economic analysis should be backed by an accurate appreciation of the environmental costs.

4.2 INTRODUCTION

Inventory is also defined as "the stock of any item or resource used in an organisation" (Chase and Aquilano 1995). Basically, types and forms of the resources used, and their consequent inventory, vary according to the area investigated, including raw materials, finished and semifinished parts, subassemblies, and supplies. In many sectors, energy is a fundamental raw material necessary for both production and services, even in its different forms, such as electricity, heat and steam, and mechanical energy. In other sectors (e.g., power stations), energy is the last product to be supplied to final or intermediate customers. Frequently, these systems convert energy from one type to another, thus behaving as the more conventional productive systems, that is, adopting stock-keeping policies, and ordering and accumulating the resources necessary to meet demand. The description given and the growing attention to energy policies underline the importance of incorporating energy-related issues to inventory principles and vice versa. (Incidentally, it is interesting to note how, as in Hudson and Badiru [2008], energy is classified into two basic types: kinetic and potential, where potential energy represents stored energy, as well as energy of position.) As far as early literature on the economic features of energy storage is concerned, the contribution by Grubbström and Hultman (1989) is to be mentioned, with particular reference to the proposed model for exergy storage of a system subject to heat transfer.

As an introductory comment, it is convenient to keep in mind one of the basic principles of inventory economics, which is based on the provisioning costs (e.g., setup and purchasing) and the costs related to stock creation and maintenance (e.g., see Winston 1994). The latter cost is generally given per unit of inventory and per time unit and it may be connected both to the opportunity costs (capital tied up to the inventory) and to the pure components of the holding activities (e.g., risks and costs of obsolescence, taxes on inventory, damages and spoilage, stockout probabilities, insurance, buildings, operating costs, etc.). In addition, purchasing costs may include, in the case of production, variable costs, labor, raw materials, and, as in the case of the present contribution, energy.

Of course, the activities related to stock energy or substances to be used for energy generation must also comply with these inventory laws and principles.

However, in the case of energy-related raw materials and energy itself, the price paid for the raw material or the pure energy (i.e., oil, gas, carbon, and, as derived energy sources, electricity and process residuals such as hot water and air) cannot be neglected either. In fact, this component may be estimated by opportunity cost and, above all, it may fluctuate widely within narrow time windows because of speculation or demand–offer mechanisms. This aspect significantly impacts modeling; it implies that the purchase price cannot be assumed as constant over time, nor can the holding cost be considered as a given and consolidated value. These two principles may be considered connected thanks to the opportunity cost mechanism. In addition, this reasoning suggests that the variable holding cost deserves further analysis, in accordance with the needs implicit in inventory-related energy issues, as discussed Section 4.5.

4.3 A PRELIMINARY SCHEME

In a general framework, the scheme proposed in the European Commission (EU) reference document (2008) may be suitable for the present case. Energy inputs into a process with a given productive capacity (e.g., tons per week) may differ in type: steam, electricity, and fuels. Once the resource requirements have been adjusted by their conversion efficiency (e.g., 35–55 percent for electricity and 85–95 percent for steam generation), it is possible to appreciate the gross requirement of energy for the production capacity. The concept may be extended to include energy recovery and different flows of products in the process. Adapting the scheme mentioned to the purpose of the present chapter, it is possible to offer the general view of a productive system shown in Figure 4.1.

$$\text{Specific Energy Content}_{\text{Product 1}} = \frac{\sum_i [E_{i,in} + E_{i,rec}] + E_{w,in} - \sum_i (E_{i,out}) - E_{w,rec}}{P_1} \quad \frac{\text{GJ}}{\text{ton}} \quad (4.1)$$

Equation 4.1 allows the appreciation of the specific energy consumption related to each unit of production to be conveniently expressed, for example, as GJ/ton or kWh/item. By focusing on a simple system producing one main product, for example, product 1, various inputs are necessary to the system so as to complete the amount P_1 of products type 1. These inputs are represented by the flows (F_j) of the raw materials (N) to be processed, and by the quantities of the different forms of energy $(E_{i,in})$ to be used in the process itself (e.g., i = steam/hot water, electricity, fuel, mechanical, other). Specific energy consumption may be defined as the net amount of energy used to produce one unit of output (in terms of liters, items, tons, etc.), taking into account the net exported energy. Depending on the system boundaries, it could be necessary to include the energy conversion factors, so as to appreciate the real gross requirement of energy (e.g., efficiency in producing steam and hot water is significantly larger than the efficiency in electricity generation, thus introducing the issue of the type of energy vector used).

As the productive system may provide both products and services, the different forms of net exported energy $(E_{i,out})$ may represent system outputs, too. These are the cases of energy amounts used in other process units, but, as a function of the boundaries considered, it is also the case of industries recovering their warm residuals

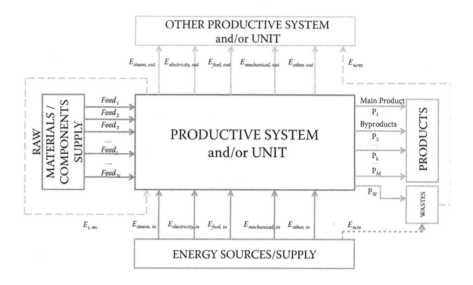

FIGURE 4.1 A general view of energy flows in a productive system (adapted from the EU 2008).

for district heating. In such situations, $E_{i,out}$ terms represent recovered resources. Of course, the production of product type 1 may also generate byproducts (as in various chemical processes; i.e., P_2, ... P_k, ... P_M), as well as wastes (in terms of residuals) which, in turn, could require energy ($E_{W,in}$) to be disposed of (e.g., incineration), eventually generating an energy export ($E_{w,rec}$). Finally, technical solutions may allow the recovery of a part of energy ($E_{i,rec}$) within the process itself. Other energy efficiency indicators may be formulated, according to investigation needs, such as definition of the system boundaries, appreciation of system-oriented or product-oriented energy requirements, multi-product environments and technical comparison of different processes.

However, the conclusive remark is that the whole set of energy contributions described is transferred into the final products; in other words, each product contains an energy contribution that also depends on the process and the type of energy used, thus making a substantial difference in terms of equivalent emissions into the environment, too. Finally, it should be noted that two types of energy contributions exist: the first is a fixed one (everything needed to support production, regardless of the quantities produced, such as lighting, maintaining vessel pressures, minimal flows in pipes, and furnace temperatures), and the second is a variable contribution (e.g., the energy spent for cutting and heating raw materials).

More extensively, it is quite interesting to note how, according to a modern view, goods are no longer evaluated mainly on the basis of their economic value or associated monetary flows, but also according to their energy requirement and content. For example, this concept is applied to buildings (Yohanis and Norton 2006), where energy is categorized into embodied and operational energy, and to photovoltaic facilities (Muneer et al. 2006). The former case suggests an interesting dualism between setup (embodied energy) and holding costs (operational energy), while the latter example gives outstanding evidence of how the two approaches (economic and energetic) may lead to controversial conclusions (the payback of the energy embodied in the facility is about eight years, while the monetary payback is estimated around ninety-five years). From a wider point of view, it can be stated that energy is a commodity encompassing each object, which, on its turn, requires (or releases) energy during its life and up to its end.

4.4 STOCKING ENERGY VECTORS

The fluctuations in the availability of energy-related raw materials, together with the volatility of their processing, have led companies and national states to organize their own network of reservoirs, buying and stocking these materials during positive periods (low prices, common availability) and consuming them during negative ones. Such an approach may also be determined by seasonal needs, as in the case of stocking fuels during the warmer period to be sufficiently protected during the winter. This is the case of strategic gas stocking and also of any traditional household that collects wood and biomass during the harvesting period to be protected for the cold season. These simple examples show how stocking materials with energy-related purposes is very similar to the classic role of inventories, which are used as flying wheels to smooth the imbalance between demand and offer, covering peaks and buffering during troughs. Energy-related stocks also play an important strategic role, preventing national traumas during seasons of political disruption (oil shocks, interrupted gas provisions).

Historically, literature has paid greatest attention to the study of oil stock, probably because of its political relevance, raw material versatility (oil is used in heating, power generation, and propulsion) and transportability, and geographical distribution (in this respect, all the other energy sources—carbon, gas, and biomass—differ significantly). However, the technological changes and infrastructure development set the basis for new opportunities for energy-related means, as in the case of gas.

Gas is generally transported by pipelines from the extraction sites to the final destination. Such a solution presents both advantages and disadvantages, as the simple comparison with traditional manufacturing might show:

1. Pipelines are "rigid" transport systems, connecting specific points of a given network: a given supplier is connected to a specific set of customers and vice versa, even if, thanks to the pipeline grid, different nodes may be reached.
2. However, supplying consists of a continuous flow, which may be preferred to discrete provisioning, where levels vary according to steps in input and continuously in output.
3. The pipeline may be critical, as sudden breakdowns (or shutdowns) of critical points may interrupt a relevant part of the flow, or even the entire flow.
4. The gas in the pipeline is a sort of work-in-progress stock, increasing its cost while moving through the pipe itself, because of the energy required to move it and the part of the infrastructure used.

From this point of view, liquid natural gas (LNG) and regassifiers represent an interesting alternative (Energy Information Administration [EIA] 2003), giving flexibility to the system. The first plant of LNG started in Arzew, Algeria, in 1964. Since then the growth has been impressive, supported by technological changes and cost reductions at the different phases of the chain (Greaker and Lund Sagen 2008). Regassifiers require gas stocking in harbors or, in general, close to the coasts. The natural resource is stocked and shipped by cooling the gas itself at extremely low temperature, thereby incurring an energy cost for the pumping at relevant pressure and for keeping temperatures at low levels (transportation takes place at –160°C). The same applies at the final destination, where ships fill stocking vessels with the transported LNG low temperatures until the gas is released into the pipeline distribution. Such a system is of great interest for its modeling and potential implications (Holz et al. 2008):

- It opens up the gas market to new suppliers who are not necessarily connected by pipelines to the final customers (several natural gas fields are far from major markets).
- Such a distribution system makes the natural gas supply chain more flexible, for example, in the choice of provisioning harbors, routing, and delivery stock points (the impact of the infrastructure is reduced, too).
- From this point of view, greater competition is stimulated and the opportunities for natural gas exploitation widened. (At present, this fuel is commonly used in power generation and is an interesting propulsion fuel, too, although its potential is limited by the scarcity of distribution points.)

However, the distribution of costs varies when compared to the traditional pipeline: a higher cost is paid for stocking (cooling and stock pressure) as an initial setup cost, while the cost of energy (distributed along the pipeline due to gas pumping) is counterbalanced by temperature preserving costs. Even if, from the 1980s, both liquefaction and regassification costs decreased significantly, it should be considered that the LNG chain is also energy intensive. About 15–20 percent of the raw material is claimed to be consumed for the entire processes (liquefaction, transport, and regassification).

Gas storage facilities are used to accumulate gas during the year and to regulate gas production, making it more stable. This approach covers strategic, economic, and technical issues (see, e.g., EIA 2006). Storage may be completed underground in exploited gas fields, aquifers, and caverns, although artificial tanks and pipelines themselves may play an integrating role balancing pressure and demand. This leads to competition over prices and contemporary smoothing of peaks in demand. For example, in Italy, demand is extremely seasonal: in winter it is roughly twice the summer figure, thus favoring the hypothesis of stockpiling during the summer and null stock at the end of the winter season. Extracted gas may be transported and stocked closer to the final user, the alternative being to avoid extraction and keeping the natural reservoir untouched. In such cases, a difference (immediately apparent to inventory researchers) emerges in the inventory holding. When gas is kept in its natural field, the holding cost is null, but if the gas is extracted and stocked prior to its final usage, the holding cost emerges as a consequence of stock maintaining (in a simple approach, the marginal cost), plus capital tied to extraction and transport. As a consequence, storage capacity is the most relevant constraint and the strategic function of gas for civil and industrial purposes suggesting the growing importance of this factor. When dealing with gas storage, a large stock capacity also prevents the need for overdimensioning pipelines and extraction sites. It also exerts a moderating action on the market during critical periods, reducing price fluctuations. An interesting and complete model may be found in Amundsen (1991). A careful— and inventory-based—evaluation of the true convenience of gas stocking is made complex by the number and variety of parameters specifically involved. Above all, taxation emerges as a critical parameter (together with price caps), being a typical leverage used with political aims to address the strategic storage aspect of gas reservoirs. These aspects are quite complex to model, including their unpredictable effect on demand and, consequently, on the inventory needed. By way of example, the price to access storage in Italy, which is still almost monopolistic (Holz et al. 2008), also includes fees for space, injection, withdrawal, movement, and strategic storage (Italian Authority for Electric Energy and Gas 2006). It is evident that the common or independent movement of any of these figures will affect the interest in gas stocking, together with any expected fluctuations in price and demand.

In conclusion, the issue of gas storage is an emerging research area, deserving of particular attention because of its inventory-related features. Of course, even greater attention has been paid in the past to a similar case, namely, petroleum reserves (e.g., see Meyer et al. 1979; Oren and Hong Wan 1986). This is the main reason why, in the present contribution, it has been preferred to discuss energy resources and vectors other than oil, which can be regarded as a well-consolidated issue. However, the great attention paid to oil, due to its strategic role and to the shocks of the 1970s (e.g., the embargo in 1973 and the Iranian crisis in 1979), is similar to the recent attention to natural gas. Of course, differences in demand, taxation, and technical solutions (related to the physical properties of the two substances) make the two cases very different. Furthermore, the problems posed by natural gas present some conceptual similarities to hydrogen concerns, even if the two orders of scale are completely different. For example, as in Kreith and West (2004):

- Liquefaction of hydrogen requires a large amount of energy.
- Storage occurs at very high pressures, with a considerable energy input.
- Its transport still presents daunting obstacles.

Nevertheless, the conceptual similarities mentioned add interest to the development of models for the optimization of the chains of these energy vectors, including the amount of energy at the generation phase, the energy requirements, and the infrastructure investments for subsequent transport and stocking.

A second category of stocks related to energy is represented by water reservoirs, generally accumulated by dams. Where it was an option, the generation of electricity by means of hydropower has always been competitive because of its costs, and it continues to be so now because of environmental awareness and attention to power generation with low CO_2 emissions. In fact, hydropower is one of the most consolidated and time-proven ways to produce energy by renewable resources, and its technological development reached a mature standard some decades ago. The management of water resources exploited in dam-based plants has already been dealt with in the literature. As in Prabhu (1980), water reservoirs allow interesting applications of continuous time inventory models, being a storage process with random input and output of finite capacity. His bibliography shows how P.A.P Moran formulated his theory on dams in 1959 and the first works date back to the 1950s (H. E. Hurst, J. Little, Tjalling C. Koopmans, J. Gani and R. Pyke), mentioning how demand may be originated both by electric energy and water to be supplied. The elegant approach adopted, based on the queuing theory, is worthy of application to the economics of inventories, considering the holding cost of water storage. In a modern view, such a cost should include:

- The cost of the structure, together with its maintenance and updating
- The costs related to the environmental impact, comprising not only the local changes (effect of dams and water storage on the site, such as soil assessment, changes in humidity and temperature, reflectivity of the artificial lake surface), but also the entire flow of the original intercepted streams
- Hidden costs related to the missed opportunities for water utilization, such as agriculture and fish farming; in particular, it should be noted that extremely dry seasons may oblige the managers to lower dam levels, i.e., spillage, to prevent irreparable damage to agriculture, industry, and other activities, thus determining an additional holding cost that should not be overlooked

In conclusion, the need for accurate models of stock management, visibly quantified by the water level, emerges as particularly relevant. Such models should help in smoothing irregular provisioning of the inflows, taking into account the risks related to future incomes (stock usage may determine a deficit due to dry seasons or a loss in capacity accumulation during a wet period and already high stock levels).

The issue of water resource utilization and, more specifically, its optimization is comprehensively discussed by Loganathan (2008) and the authors wish to acknowledge his contribution.

When the specific issue of energy is taken into account, various technical solutions to its generation (e.g., fossil fuels, nuclear, hydro, etc.) present their own specific features (Couto de Oliveira et al. 2002), which are differently, though significantly, influenced by the management policies in the stock of the energy generation source. For example, operating costs are mainly determined by fuels in fossil-based power stations, while equipment construction and environmental impact mainly affect hydro and nuclear power plants. In particular, water reservoirs, when freed to be turbined for electricity generation (outflow), might become the reservoir of the subsequent power station (inflow), like serial inventories. In addition, the decision to produce "clean" energy by water saves the larger operating costs of the alternative fossil-fuel power stations, but reduces the stock of water in the reservoir, too; that is, the production capacity for the future is reduced in favor of a present use. Of course, the problem of the relative influence of immediate cost functions on future ones is perceived differently, according to the sensitivity of the system to the scarcity of water, as in small capacity hydro networks and extreme water fluctuations from season to season.

The appropriate modeling for these systems may vary (e.g., stochastic dual dynamic programming proposed by Pereira and Pinto [1991] and Read and George [1990], or simulation-based as in Read and George [1990]), since decisions and the consequent optimization depend on the forecast in inflows.

Finally, underground thermal energy storage (UTES) technology should be mentioned. Such a technology implies the storage of energy wastes and, typically, solar thermal energy in reservoirs for future use (Sanner et al. 2003). The convenience of such a technique depends on the technical solution adopted (aquifers, boreholes and pipes, and cavern storage), which offers different equipment flexibility and investments (setup cost). Furthermore, holding costs are related to the cost of the depreciation of the dedicated buildings, plus the decay in the energy content, so the energy level of the stock held must be maintained to prevent its value per unit of time from decreasing over time, also depending on the demand profile, which determines when the stocked energy will be used. Variable costs are determined by the energy used to move fluids from and to the reservoirs. In fact, thermal storage may occur on different time horizons, from short-term utilization up to seasonal. The UTES technical solution and the underlying problems leave plenty of room for optimization, which could be carried out according to inventory principles, especially because of the developing research and market interest in high temperature (>50°C) heat storage. Nevertheless, underground storage goes beyond the case of thermal energy. For example, excess energy may be stored by compressing air and pumping it into salt caverns, abandoned mines, and depleted gas fields or buried pipes. An alterative, as in hydro plants, is pumping water into higher basins for subsequent turbining (Patel 2006). The high storage capacity of the compressed air solution is also associated with relevant investments; for example, due to the problem of cooling the compressed air, the cost of the power system varies between $1000–$1500 per kilowatt). However, storage capacity may be a strategic element of choice when positioning new wind farms, enabling energy to be stored when wind power exceeds electricity demand. In fact, the development of renewable sources, such as wind and solar generation, focuses attention on the intermittency implicit in these types of resources. As penetration levels increase, finding a balance between fluctuating renewable generation and fluctuating demand becomes more difficult and more expensive to manage (Moutoux and Barnes 2007). Solutions to mitigating the intermittency of wind generation by storage are, in addition to the cases already mentioned, conventional batteries, flywheels, ultracapacitors, flow batteries, and superconducting magnetic energy storage. A complex case is described in Denholm (2006), where a completely renewable generation system combines wind energy, compressed air energy storage (CAES), and biomass gasification. Baseload wind systems using CAES may require combustion of natural gas and the development of long distance transmission, thus decreasing the environmental and social compatibility of the system. However, the use of biofuel might reduce the net carbon emissions of the system, increasing its renewable features. CAES systems are also mentioned as an emerging technique (EU 2008), because of their ability in managing peaks of demand and offering opportunities for efficiency improvement (Najjar and Jubeh 2006).

UTES is generally intended as energy stocking for the colder period; however, stocking for warmer periods may also be practiced. The basic question is that demand for cooling buildings has significantly increased in recent years due to several factors, such as population and rising comfort standards and, not least, global warming. The most common equipment for building cooling is electrically driven devices. A significantly less energy-consuming alternative is to stock "winter cold" in the form of snow and ice for subsequent utilization during the summer. This ancient technique is feasible in large areas of the world. A large-scale application is the snow cooling plant at the hospital in Sundsvall, Sweden, operating since 2000 (Skogsberg and Nordell 2001). Snow is stored in a shallow pit (140 × 60 m) of watertight asphalt, with a capacity for 60,000 m³ (40,000 tons) of snow. Meltwater from snow storage is cleaned and pumped to the hospital and, after the cooling usage, the heated meltwater is recirculated into the snow storage. The average cooling energy required by the hospital during the summer amounts to approximately 700 MWh, with a maximum cooling power of 1.4 MW; 93 percent of the cooling demand was attained by 19,000 m³ of snow and the rest by a cooling machine. Most of the snow (75 percent) was natural, integrated as needed by snowguns. Without thermal insulation and cold extraction, a 30,000 m³ snow pile in Sundsvall would be gone by mid-June. A 0.2 m cover of wood chips reduces melting considerably; in this case, 75 percent of the snow is preserved and may be utilized for the energy saving described, which, in conclusion, is characterized by an extremely cheap cost of energy (average close to 60 €/MWh) and a long life expectancy (40 years). Moreover, snow/ice cooling plants have a great potential in industry, agriculture, and comfort cooling; artificial cooling media are often inflammable, poisonous, and environmentally hazardous.

In conclusion, the problem of energy stocking is strategic and is gaining increased importance with renewable energies (e.g., stocking solar heat during the day for night usage or the presence of wind during a period of no demand). The issue is also addressed by the EU reference document on energy efficiency techniques (2008). As forms of energy are very different, some of them are more difficult or impossible to store. Mechanical and thermal energy cannot be easily stored (e.g., dams and hot water tanks). However, the most important case is electricity. Potentially, electricity is stored in batteries, which is a form of chemical storage, just like fuels. This probably represents the most successful form of energy storage in terms of transportability, energy density, and duration. In conclusion, the advantages of storage technologies are obvious, but they also offer strategic potential to improve both economics and the environmental performance of renewable energy sources; external fuels are no longer necessary and excess generation by renewables may be efficiently used (Gross et al. 2003).

4.5 HOLDING COSTS AND ENERGY

Variable holding cost is a research issue that has attracted attention for years. Works may be found in the literature in 1980 (Muhlemann and Valtis-Spanopoulos 1980) and 1982 (Weiss 1982), while a comprehensive review on perishable good inventories may be found in Nahmias (1982). There are further excellent investigations dating from the 1990s (Goh 1994; Giri and Chaudhuri 1998) and more recently by Alfares (2007) and Ferguson et al. (2007). The contribution by Goh (1994) is particularly valuable because of the two types of holding cost functions discussed: (1) the nonlinear function of the period spent in inventory by the item, and (2) the nonlinear function related to the amount of on-hand inventory. As far as the link between variable holding cost and this chapter topic is concerned, it is worth noting that variable holding costs may depend on the variability of the economic value of the item and on the energy required to preserve it. This is the case of food, where the energy and the relevant technical equipment required to preserve it is largely influenced by external conditions (temperature, humidity, etc.), thus depending on season and geographic stocking of the products, as in the case of the different levels of energy needed to preserve food in winter and summer.

Just to mention a couple of familiar examples, freezers and refrigerators are commonly used, together with the consequent problem of energy bills and the decision whether to keep things fresh in the refrigerator or pretreat and freeze them. Energy is necessary to preserve food and other products such as drugs, but also blood; the greater the energy used (frozen food vs. fresh), the longer the life of the product. Another case of increased holding costs is related to the heating power; a biomass subjected to compression (pellet and briquettes production) increases its energy content per unit of weight, thus increasing its commercial value and, therefore, its holding cost. This topic will be discussed in Section 4.8. The holding cost of biomass derivatives increases as a consequence of the manufacturing process added value and the cost of storage away from moisture. In other words, the relative values of the holding costs introduce the dilemma about the timing and the opportunity of transforming the biomass into its derivatives. However, the value of the item stored may also decrease over time, for example, because of its energy content decay or because of the energy spent to preserve the stock itself. The former is the case of batteries, the charge of which decreases over time, and the second is the case of regassifiers and UTES.

The work by Alfares (2007) introduces a holding cost variable according to a step function, this being an interesting model for value increase of biomass because of pellet or briquette manufacturing or, in general, for products for which longer storage periods require extra care. The same contribution (Alfares, 2007) emphasizes how the model proposed problems, where an item is stored in different places with different holding costs (e.g., perishable items, for which the longer the time to be spent in storage, the more sophisticated the services and facilities needed and, consequently, the higher the holding costs). Also Giri and Chaudhuri (1994) remark on how variable holding costs are related to supermarket management of fresh food: increasing costs are determined by improved storage facilities to prevent spoilage and maintain freshness. Of course, integrating policies other than preservation may be put into action, such as markdowns or removal of spoiled products. (See also Ferguson et al. [2007] for a real-life application of variable holding cost models.)

However, as technological and strategic issues drastically influence the matter in question, it is not possible to neglect the impact of alternatives in the supply chain management. An example is offered by Bendiksen and Dreyer (2003) referring to fish processing in Norway. Even if the focus of the study is on plant locations (of a specific region and product) and the relative value, the impact of the technological change related to the supply chain of the product considered is evident, as it may bring into question the strategic geographic position of the plant itself. The possibility of storing fish on board immediately after the open-sea capture without quality loss might change the structure of the chain and the consequent distribution of inventories. Time for open-sea fishing is extended by onboard storing. The most relevant technological change exists on new vessels, where the catch is frozen on board, thus changing the supply to processing plants from fresh to frozen raw material. Such a change determines the need for new storing plants for frozen fish (twelve new plants in five–six years, according to the Norwegian case) and processing cycles. In addition, the technological change modifies the approach to the strategic location of fish processing plants, together with the role of the supply-chain actors. Closeness to customer areas or fast transport stations becomes more remunerative than closeness to rich fishing areas. What is relevant to our purpose is that such a case also shows an example of a different distribution of inventories obtained by putting a larger energy content at the preliminary stage (cooler fish storage conditions for longer periods). This is not a new approach in the food sector; for example, cheese and meat may be seasoned or smoked for longer conservation, but such a practice requires longer times and energy contributions (smoking, drying buildings, ventilation, temperature and humidity control). When compared to fresher products and regardless of the taste, the difference lies in higher prices and longer shelf lives. Nevertheless, the emerging concept is that at the preliminary stages, the product acquires higher values because of the contribution of energy (frozen stage and storage). Fluctuations in market demand may be covered thanks to a larger inventory level and a longer product life (frozen). The case discussed introduces the topic of "cold logistic chains," which are significantly affected by energy-related issues.

4.6 "COLD" AND "FRESH" SUPPLY CHAINS

Heating and freezing are some of the oldest physical processes for food conservation, and the contribution of energy to the processes themselves is evident. Certainly, heating was formerly used because it was easy to heat by fire; however producing cold temperatures, excluding places where ice naturally forms, turned out to be a complicated task for centuries. In the nineteenth century, Norway conducted a flourishing trade in ice and transported it as far as the tropics. The first cold-storage premises were opened in Chicago in 1878, almost contemporarily to the first consignments of frozen meat from South America and Australia to Europe. The first household refrigerators appeared on the market in 1910. In 1925, trade in frozen goods for the home began in the United States. Refrigerators became widespread from 1950 onward and freezers from 1960 (Geiges 1996). Nowadays, chilled and frozen foods are moved worldwide. The different transport chains, together with their technical features, are described in James et al. (2006). Along these supply chains, energy plays a strategic role, being necessary to guarantee quality-based processes, influencing the economic performance, too. For example, supermarket refrigeration systems consume about 20 percent of the total energy of the store. This figure must be corrected to also take into account the auxiliary equipment consumption (14 percent) and the energy used for the building air-conditioning (17 percent; Getu and Basal 2007).

However, cold supply chain management implies a rather peculiar approach. An interesting introduction to the problem may be found in Bogataj et al. (2005). The emerging factor in cold supply chains is the visibility and the control, in particular of the temperature at each step of the chain, which is necessary to preserve food and guarantee its quality to final customers (these issues easily extend to perishables). Of course, the progress in technologies and logistics has boosted the cold chains, thanks to monitoring, efficient transportation, and communication information technologies. It should be noted that temperature control is a critical issue; temperatures higher than the set values may definitely compromise the product, while lower ones may compromise the quality. According to Bogataj et al. (2005), this issue is also emerging thanks to regulations such as EN92/1CEE on temperature recording in transportation, stocking, and selling points, giving us the modern concept of "cold traceability." In their contribution, starting from Grubbstrom's (1967, 1998) approach to net present value (NPV) and MRP, Bogataj et al. (2005) argue that the analysis in frequency space may be more effective than the traditional time-based one, offering the opportunity to model perturbations in temperature and the necessary prompt reactions to them. The variations in temperature (disturbances) are introduced as delays determining an impact on product deterioration, therefore adding an additional cost due to planned cooling. The NPV analysis allows the appreciation of the impact of energy-related costs in the inventories distributed over the supply chain and the entire time horizon. Such an approach may allow the understanding of the basic difference between the fresh product chain (shorter product lives, fast transportation, lower energy contribution) and the frozen product chain (longer product lives, slower transportation, higher energy contribution). The two chains may be compared, remembering that mixed situations may be found in practice (e.g., freezing–defrosting–cooking–freezing or fresh delivery–freezing).

The study of food supply chains is of challenging interest to inventory and production researchers. In fact, the fresh food chain is affected by peculiarities close to typical production issues, but worth investigating (e.g., Akkerman and Van Donk 2007):

1. Frequently, setup times are sequence dependent (think about cleaning, equipment regulation, and selection of packages) both in time and costs.
2. Storage should be limited for fresh food, not only because of perishability and shelf life, but also because of dedicated equipment and space, which could be additionally limited when product mixing is to be avoided.
3. Processing should be fast, by means of traced and high-quality systems.

4.7 ENERGY-RELATED PRACTICES: IMPLICATIONS ON INVENTORY MANAGEMENT

It has been shown how the contribution of energy to productive costs may be particularly relevant. However, it could even be overwhelming in specific industrial sectors, such as the production of silica bars, steel, and primary aluminum. This has led practitioners to focus on technical solutions allowing the reduction of energy consumption per unit produced and energy waste, so as to gain competitive advantages. Frequently, such technical solutions impact inventory features, stocked quantities, and their management. The following section will illustrate some of these cases to clarify the problem and note its impact on inventory management.

4.7.1 THE ALUMINUM CASE

Aluminum is widely recognized as the "green metal," because of its ease of recycling, durability, and properties, such as lightness and resistance to corrosion. The productive cycle of aluminum and its alloys is based on minerals (primary industry, using bauxite) or recycled scraps (secondary industry), the latter process being several times less energy intensive. Recycling processes, as in the case of steel and copper, significantly contribute to the sustainable development of society; metals are continuously reused at the end of product lifetimes. However, the energy spent on melting and refining the material is definitely lost because of metal solidification in ingots. Therefore, an approximately identical quantity of energy is to be spent if the material is subsequently reused for casting or die-casting processes, generally leading to cost and pollution inefficiencies from a chain perspective. This has generated interest in alternative structures of supply chains, as described in Diana et al. (2007). The traditional aluminum supply chain is also described and studied in Khoo et al. (2001), where a conventional supply of solid ingots is considered. An alternative practice was initially observed in Germany and is spreading in Italy. It is linked to the supply of the molten metal, transported by adequately equipped trucks carrying coated and insulated ladles. Of course, this supply method enables the final user, when properly equipped with cranes, to save the remelting energy and reduce the storage costs (as fewer ingots are stored in warehouses). The refiner gets some advantages, too: the productive cycle of furnaces may be optimized to reduce energy consumption and storage costs are also reduced. Of course, there are some disadvantages, especially from the managerial point of view:

- The necessary closer cooperation between the user and the supplier is not always perceived as a competitive advantage, particularly on the buyer's side.
- Both melting (supplier) and production (buyer) cycles have to be synchronized to prevent dead times in the buyer's production. This may occur when trucks are late or when production is delayed even though the molten metal, contained in the transported ladle, may be kept in the liquid phase for a given time with limited energy contributions. Such a requirement makes the system more rigid from the managerial point of view.
- The limitation is particularly apparent with uncertain delivery times, considering the truck-based transport system is more prone to delays.
- The quantity of material transported by the truck is less in the case of liquid metal than in the case of solid ingots because of the ladle shape and weight.
- Investment in new equipment (ladles, cranes, fixtures) is needed.

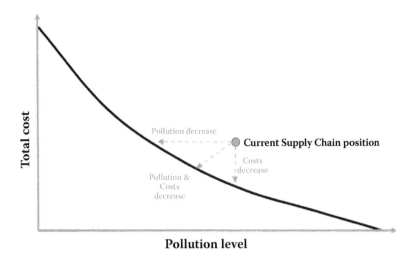

FIGURE 4.2 Schematic representation of the Pareto efficient frontier.

These considerations represent a Pareto efficient frontier, which can be identified by analytical models (Diana et al. 2007). Such an efficient frontier is of particular interest when environmental concerns are considered, too, as emissions saved at the remelting phase are partially lost by the reduced transport capacity. A schematic representation of the Pareto efficient frontier is offered in Figure 4.2. In addition, a mix of different types of supplies (i.e., sometimes metal ingots and other times molten metal ladles) may be a strategic leverage to mitigate the impact on the productive system flexibility.

From the inventory point of view, this practice leads to a decrease in the final product stocks at the supplier, which induces a reduction in stocked raw material at the buyer. These aspects are highlighted by certain cost components, such as:

- Aluminum ingot solidification costs, which cover the time for which a chosen machine (and the related space) is dedicated to containing the molten metal during the transition from liquid to solid phase, plus the time needed to cool the solid ingot before safe handling.
- Yearly depreciation of the specialist equipment (e.g., portable ladles to carry out transportation and overhead cranes with related fixtures).
- Ingot holding costs, which include the space required as well as the inventory equipment, plus management and financial costs (capital tied up and opportunity costs).
- Costs due for ingot melting at the buyer's site.
- Warm-up costs of the liquid aluminum from the transported ladle temperature to the furnace temperature and for maintaining the temperature at the ladle before use.

The case described presents some similarities with steel and iron production.

4.7.2 THE IRON AND STEEL CASE

The iron and steel industry is one of the most strategic and consolidated, but it is also one of the most energy intensive processes, accounting for the 19 percent of the European industrial energy use in 2004. A common plant type consists of an electric arc furnace (EAF) feeding a continuous casting line, the final products of which are square-section billets. Given the high temperature of the process (about 1500°C), billet cooling requires time before delivery to the final customer. Such a productive phase takes time and so it is susceptible to optimization, especially because of holding costs for the "warm inventory," which are directly proportional to the time interval during which a billet remains in the warehouse, even if it is simply cooling (Zanoni and Zavanella 2005; Ferretti et al. 2006). A survey on steel plant management (Tang et al. 2001) emphasized that the optimization may refer to the whole system or to only a part of it, as typical stages of steel production are steel making (SM), continuous casting (CC), and hot rolling (HR). In this case, when the billets are produced for a subsequent rolling phase, the loss of energy related to their cooling is evident. In fact, such a situation implies that billets are cooled, warmed, again, and then rolled. Cooling requires time and space, while warming needs a dedicated furnace and the related energy contribution. Alternative technologies, such as compact strip production (CSP), present a higher energy performance and a simplification in the process stages. Basically, slab continuous casting is coupled with continuous strip rolling, thus implying the on-time utilization of cast material, shortly after its exit from the EAF. A limiting process is represented by endless strip production (ESP), which will allow producing strips directly from liquid-core steel. Savings in holding costs (space, cooling, and stocking times) and the reduction of the work in process (WIP) are evident. These savings are to be added to the lower capital tied up with slabs, and the lower emission levels connected with suppressing the reheating process.

The viability of using the energy content transferred to the billet during the SM phase, immediately after the CC, has also been exploited by transporting billets in special sarcophagi made by refractory. It is reasonable to assume that the same considerations made for the aluminum case still apply: the two production systems (SM plus CC and HR) must be synchronized, additional equipment is required, and, in the case of truck transport, capacity per delivery is reduced.

It is worth noting that such a practice is recommended in the EU (2001) best available technologies (BATs). In particular, the reduction of heat loss is recommended for intermediate products, thus impacting WIP storage, logistics, and product flows. Minimizing storage time and insulating the slabs/blooms (e.g., by heat conservation boxes or thermal covers) is suggested. Furthermore, changes in logistics and intermediate storage lead to the maximization of hot charging, up to direct charging or rolling.

Both of the cases described show how inventories may give up one of their traditional roles—the function of decoupling (acting as reservoirs or buffers) two productive stages or subsequent levels of supply chains. Such a change is dictated by energy-related issues, which, in their turn, impact on convenience (cost compression), sustainability (emission reductions), technological changes, and management principles (scheduling, WIP, and lead-times evaluation).

4.8 BIOMASS AND ITS DERIVATIVES

Recent years have seen a renewed interest in biomass, together with a growth in its transformation processes and the use of its derivates used as fuels. Such an interest is linked both to environmental concerns (neutral cycle of biomasses, which act as CO_2 traps during their lives, releasing it because of usage or decomposition) and the cost increase of traditional materials for energy generation (e.g., oil and gas). Basically, biomass can be regarded as sun energy stored to be further extracted by different processes, such as wood burning. Biomass represents the first type of fuel used by mankind, but its prospective in supply and storage are quite new, given the peculiar aspects of its more recent applications. This fact is probably twofold. First, new biomass processes are flourishing and becoming competitive. Second, biomass has usually been considered as a local resource, even though the demand and the available sources are not homogeneously spread all over the world. Some countries have excess biomass for their own energy economy, while in other countries, biomass availability is limited due to population density and industrial activity (Forsberg 2000). Furthermore, the increasing use of bioenergy in Europe has already created a new market for the supply of this resource, which can be satisfied (in the short term) also by an increase in the access to biomass over a larger geographical area, thus increasing the interest for the relevant supply chains, particularly processes, storage, and transportation. In other terms, using an appropriate logistic system, access to biomass can be improved over a large geographical area (even though some countries, such as Italy, explicitly encourage "short" [70 km] biomass supply chains). Both the studies by Forsberg (2000) and Hamelinck et al. (2005), using life cycle inventory (LCI), confirm that long-range transportation may also allow an efficient exploitation of biomass, without losing its environmental benefits.

However, as the authors suggest, the optimization of the chain should consider the processes occurring at the different levels of the chain itself. In fact, technological evolution enables biomass transformation to facilitate its storage and transport along the chain up to its final use (examples of derivatives are biofuels, pellets, briquettes, and chips). These transformations require energy utilization, which basically determines the increase of the specific heat (i.e., kilojoules per kilogram or kilojoules per cubic meter) and the density of the biomass, opening interesting frontiers to research applied to the optimization of the biomass supply chain and its inventories. For example, Gigler et al. (1999) discuss the configuration of biomass supply (willows) to energy plants. Basically, wood may be supplied to plants differing in power size and material utilization (combustion or gasification), but alternatives characterize the chain steps. Wood may be stocked as chips, dried (forced cycles), and finally delivered after a given time span (the typical horizon is from a few months to one year). Alternatively, willows may be simply cut (chunks or stems), transported, and centrally chipped. According to Gigler et al., the former alternative is more economically attractive for deliveries within six months, while the latter applies to longer-term deliveries. A significant impact is exerted by drying, which can be energy based (forced drying) or natural, thus determining an impact on costs, emissions, and energy requirement by the cycle. A further interesting finding is that supply chains for gasification plants are more expensive than conventional combustion ones if the cost evaluation is expressed

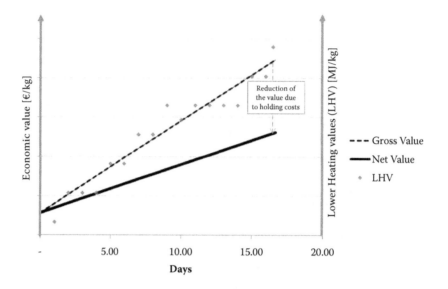

FIGURE 4.3 Experimental investigation on the lower heating value of wood chips (poplar) during natural drying, and estimated gross and net values.

in euro per ton dry matter (DM), but the cost is lower when expressed as euro per kilowatt hour, due to the higher efficiency of the technological cycle. This shows the appropriateness of a nontraditional analysis, which does not take into account solely conventional costs (e.g., euros per ton or euros per year), but also engineering-oriented measures such as euros per kilowatt hour or euros per $kgCO_2$ to consider emission costs. For example, Figure 4.3 shows the trend of the lower heating value (LHV), measured on field, of wood residuals (chips) in the sawmill department of an Italian company. As time flows, natural drying lets LHV increase, thus resulting in an increased chip value as fuel. However, holding costs mitigate this positive effect, leading to a lower net value, which depends on the stocking equipment.

The distribution of pellets represents a very interesting reference, too. According to Vinterback (2004), distribution may represent 30 percent of the cost for a residential pellet consumer, and inventory may be greater than 50 percent. Sensitivity to moisture by the pellet fuel makes storage a rather critical activity. Improving pellet distribution and, in particular, inventory management at the distributors might point users toward this heating alternative, appropriately considered as environmentally friendly. Technological innovation has considered storage at the final customer by the producer—even if demand profile and productive capacity generally make inventory a necessary solution—and room control during pellet storing, even at the transportation phase. The ordering mechanism is a rather interesting situation for inventory modeling; there exists a reordering point, a maximum storage capacity, and the need for a safety stock and uncertain demand.

It is worth noting that wood supply chains are complicated by the profiles of demand and raw material availability, the latter being related to the seasonality of forest management. This case is frequent in the pulp industry (Carlsson and Ronnqvist 2005): the availability of raw material naturally increases in the forest up to midsummer, while harvesting builds up the stock between autumn and early spring, depending on both cultivating traditions and regulations. In the case of pulp, demand from mills is considered stable while, in the case of biomass fuels, demand is affected by high seasonality, so that the need for balancing raw material availability, production capacity, and final demand reflects on a series of inventories distributed over the supply chain.

In general, the design of a biomass supply chain is particularly interesting because of the transformations of the products (which substantially change their technical parameters) and the impact of transportation on optimal choices (Hamelinck et al. 2005; Gronalt and Rauch 2007). It is also accepted that, within the renewable resources framework, biomasses present a complicated scenario, because of the variety of feedstocks, conversion technologies, and final usage (e.g., domestic or industrial, electricity- and heat-oriented ones). Therefore, a key issue is to secure sustainable supplies of relatively low cost fuels (Gross et al. 2003). However, their benefits are evident, being largely available, flexible (as described), and environmentally friendly. As shown in Hamelinck et al. 2005, Gronalt and Rauch 2007, and Gross et al. 2003, biomass presents interesting features for its alternatives in supply chains and the related inventories. For example, demand is affected by seasonal fluctuations to be added to "regular consumption (the latter component being significantly determined by the paper and furniture industries and/or the combined heat and power plants, producing both heat and electricity)." As far as transportation is concerned, biomass in loose form is competitive within a short range (50 km) and wood chipping is effective in intermediate ranges (up to 100 km), while longer distances require biomass transformation into derivatives (e.g., pellets, biofuels). Safety stocks must be kept at terminals and plants (expressed in millions of loose cubic meters), guaranteeing a set service level in a variable lead time context, limited stock capacity, and seasonal availability of raw material. Harvesting windows are limited, according to regulations and selected species, and the energetic properties of the raw materials are also highly variable, as in the case of wet springs and snowy winters and vice versa, thus explaining why pretreatments are frequently hoped for. Furthermore, harvesting methods may differ and transformation processes, even by mobile equipment, may take place at different levels/steps of the chain, altering biomass features and introducing energy contributions.

These observations lead to attention on conversion processes, which differ in technology cost and environmental impact (e.g., see Hamelinck et al. 2005), but also offer to assimilate, from the distribution point of view, biofuels (such as refined ones) to fossil fuels. Basically, bioenergy may be transported in different forms (chips, pellets, charcoal, ethanol, etc., and, of course, electricity). Such bioenergy chains, including their different products and densification processes—aimed at improving product technical features and reducing transport movements and volumes occupied at the storage stages—differ in external energy input and energy density of the output. Furthermore, they also differ in vulnerability to, for example, moisture (drying may be natural or artificial, but ineffective storage may thwart the process), bacteria, accidents (such as material losses for storage activities; about 15 percent), and self-ignition. It is interesting to note that biomass deterioration is mainly dependent on the material properties, such as moisture content, available nutrients, and chemical structure. It also depends on the type of storage (Jirjis 1995): for example, the low-moisture content of pellets limits the growth of microorganisms (Lehtikangas 2000).

Inventory models encompass these aspects in the concept of holding costs, including obsolescence and perishability. In addition, changes in heating value suggest opportune modeling of these parameters familiar to inventory researchers and practitioners. According to Hamelinck et al. (2005), storing in open-air costs 1.1 €/(m³y), while pellets and dried chips must be bunkered at 87 €/(m³y), up to liquid storage costs at 310 €/(m³y). However, it should be emphasized that biomass inventories (and, more generally, energy-related ones) should be optimized not only according to cost function, but also according to CO_2 balance calculations, this being a rather critical aspect to sustainability. Although the environmental savings with respect to fossil fuels are evident (electricity produced by pellets is in the 7–32 kg_{CO2}/MWh range against the 370–1200 kg_{CO2}/MWh range of gas or coal), some renewable technologies should still increase their competitiveness with fossil fuels, thanks to environmental external elements. To this end, it is opportune to point out that technology changes may modify the perspective, as in the case of carbon storage in soil and litter. Furthermore, both environmental and economic balances should take into account that biomass may generate "coupled products" (e.g., glycerine from the production of rape methyl ester), thus providing additional usefulness to these natural resources (Kaltschmitt et al. 1997). Finally, CO_2 emissions and energy required in biomass supply chains severely influence the chain sustainability. The ratio between the primary energy and the delivered energy is around 1.25 for European crop pellets (the ratio rises to 1.87–1.93 for methanol chains), that is, extra energy inputs to the chain (mainly derived by fossil fuels) and material losses represent more than one-quarter of the final energy produced (the pellet chain loses about 15 percent of the initial energy). Similarly, Cocco (2007) compares three bioenergy chains, offering evidence particularly favorable to electrical power generation from biomass (miscanthus and poplar) both in terms of net balances and ratio between output and input energy. Within the framework of the assumptions made (e.g., average yield per hectare and irrigation requirement), it is interesting to observe how energy inputs flow through the chain.

4.9 CONCLUDING REMARKS

Energy belongs to and surrounds everyone's life and, probably, both increased consciousness and repeated shocks have increased the awareness of this. Inventory theory, thanks to its principles and rationale, may substantially help the understanding and the efficient use of this resource. To this end, the present chapter aimed to show how energy is related to stocks and how it may influence them. Industrial cases and examples from supply chains were presented to let the relevant aspects of the inventories, which are related to energy, emerge. The environmental topic was also addressed, this being directly linked to energy in terms of emissions and environmental accounting. The whole set of issues proposed shows how the energy-related issues are still to be investigated in the view of inventory theory, opening wide opportunities to researchers and practitioners, with a consequent benefit to the society.

REFERENCES

Akkerman, R., and D. P. van Donk. 2007. Product prioritisation in a two-stage food production system with intermediate storage. *Int J of Prod Econ* 108(1-2):43–53.

Alfares, H. K. 2007. Inventory model with stock-level dependent demand rate and variable holding cost. *Int J of Prod Econ* 108(1-2):259–265.

Amundsen, E. S. 1991. Seasonal fluctuations of demand and optimal inventories of a non-renewable resource such as natural gas. *Resour Energ* 13(3):285–306.

Bendiksen, B. J., and B. Dreyer. 2003. Technological changes—the impact on the raw material flow and production. *Eur J Oper Res* 144(2):237–246.

Bogataj, M., L. Bogataj, and R. Vodopivec. 2005. Stability of perishable goods in cold logistic chains. *Int J of Prod Econ* 93-94:345–356.

Carlsson, D., and M. Ronnqvist. 2005. Supply chain management in forestry—case studies at Sodra Cell AB. *Eur J Oper Res* 163(3):589–616.

Chase, R. B., and N. J. Aquilano. 1995. *Production and operations management: Manufacturing and services,* 7th ed. New York: McGraw-Hill/Irwin.

Cocco, D. 2007. Comparative study on energy sustainability of biofuel production chains. *Proc IME J Power Energ* 221(5):637–645.

Couto de Oliveira, G., S. Granville, and M. Pereira. 2002. Energy. In *Handbook of applied optimisation*, edited by P.M. Pardalos and M.G.C. Resende. Oxford: Oxford University Press.

Denholm, P. 2006. Improving the technical, environmental, and social performance of wind energy systems using biomass-based energy storage. *Renew Energ* 31(9):1355–1370.

Diana, A., I. Ferretti, S. Zanoni, and L. Zavanella. 2007. Greening the aluminium supply chain. *Int J of Prod Econ* 108(1-2):236–245.

Energy Information Administration (EIA). 2003. The global liquefied natural gas market: Status and outlook. DOE/EIA-0637.

Energy Information Administration (EIA). 2006. US underground natural gas storage development: 1998–2005.

European Commission (EU). December 2001. Reference document on best available techniques in the ferrous metals processing industry.

European Commission (EU). June 2008. Reference document on best available techniques for energy efficiency.

Ferguson, M., V. Jayaraman, and G. C. Souza. 2007. An application of the EOQ model with nonlinear holding cost to inventory management of perishables. *Eur J Oper Res* 180(1):485–490.

Ferretti, I., S. Zanoni, and L. Zavanella. 2006. Production-inventory scheduling using ant system metaheuristic. *Int J of Prod Econ* 104(2):317–326.

Forsberg, G. 2000. Biomass energy transport: Analysis of bioenergy transport chains using life cycle inventory method. *Biomass Bioenergy* 19(1):17–30.

Geiges, O. 1996. Microbial processes in frozen food. *Adv Space Res* 18(12):109–118.

Getu, H. M., and P. K. Basal. 2007. Modeling and performance analyses of evaporators in frozen-food supermarket display cabinets at low temperatures. *Int J Refrigeration* 30(7):1227–1243.

Gigler, J. K., G. Meerdink, E. M. T. Hendrix. 1999. Willow supply strategies to energy plants. *Biomass Bioenergy* 17(3):185–198.

Giri, B. C., and K. S. Chaudhuri. 1998. Deterministic model of perishable inventory with stock-dependent demand rate and non linear holding cost. *Eur J Oper Res* 105(3): 467–474.

Goh, M. 1994. EOQ models with general demand and holding cost functions. *Eur J Oper Res* 73(1):50–54.

Greaker, M., and E. Lund Sagen. 2008. Explaining experience curves for new energy technologies: A case study of liquefied natural gas. *Energ Econ* 30(6):2899–2911.

Gronalt, M., and P. Rauch. 2007. Designing a regional forest fuel supply network. *Biomass Bioenergy* 31(6):393–402.

Gross, R., M. Leach, and A. Bauen. 2003. Progress in renewable energy. *Environ Int* 29(1):105–122.

Grubbström, R. W. 1967. On the application of the Laplace transform in certain economic problems. *Manag Sci* 13(7):558–567.

Grubbström, R. W. 1998. A net present value approach to safety stocks in planned production. *Int J of Prod Econ* 56-57:213–229.

Grubbström, R. W. and Hultman, P. 1989. Exergetic and inventory-theoretic aspects of energy storage. *Eng Costs Prod Econ* 15, 343–350.

Hamelinck, C. N., R. A. A. Suurs, and A. Faaij. 2005. International bioenergy transport costs and energy balance. *Biomass Bioenergy* 29(4):114–134.

Holz, F., von Hirschhausen, C. and Kemfert, C. 2008. A strategic model of European gas supply (GASMOD). *Energy Economics*, 30:766–788.

Hudson, C. R., and A. B. Badiru. 2008. Energy systems. In *Operations research and management science handbook,* edited by R. A. Ravindran. Boca Raton, FL: CRC Press.

Italian Authority for Electric Energy and Gas (AEEG). 2006. Resolution #56/06. *Approvazione dei corrispettivi di impresa e determinazione dei corrispettivi unici per l'attività di stoccaggio, anno termico 2006–2007.*

James, S. J., C. James, and J. A. Evans, 2006. Modelling of food transportation systems—a review. *Int J Refrigeration* 29(6):947–957.

Jirjis, R. 1995. Storage and drying of wood fuel. *Biomass Bioenergy* 9(1-5):181–190.

Kaltschmitt, M., G. A. Reinhardt, and T. Stelzer. 1997. Life cycle analysis of biofuels under different environmental aspects. *Biomass Bioenergy* 12(2):121–134.

Khoo, H., T. A. Spedding, I. Bainbridge, and D. M. R. Taplin. 2001. Creating a green supply chain. *Greener Manag Int* 35:71–88.

Kreith, F., and R. West. 2004. Fallacies of hydrogen economy: A critical analysis of hydrogen production and utilisation. *J Energ Resour Tech* 126(4):249–257.

Loganathan, G. V. 2008. Water resources. In *Operations research and management science handbook,* edited by R. A. Ravindran. Boca Raton, FL: CRC Press.

Lehtikangas, P. 2000. Storage effects on pelletised sawdust, logging residues and bark. *Biomass Bioenergy* 19(5):287–293.

Meyer, R. R., M. H. Rothkopf, and S. A. Smith. 1979. Reliability and inventory in a production-storage systems. *Manag Sci* 25(8):799–807.

Moutoux, R., and F. Barnes. 2007. Wind integrated compressed air energy storage in Colorado. sixth biennial conference on Electrical Energy Storage Systems Applications and Technologies San Francisco, California, Sept. 23–26, 2007.

Muhlemann, A. P., and N. P. Valtis-Spanopoulos. 1980. A variable holding cost rate EOQ model. *Eur J Oper Res* 4(2):132–135.

Muneer, T., S. Younes, N. Lambert, and J. Kubie. 2006. Life cycle assessment of a medium-sized photovoltaic facility at a high latitude location. *Proc IME J Power Energ* 220(6):517–524.

Nahmias, S. 1982. Perishable inventory theory: A review. *Oper Res* 30(4):680–708.

Najjar, Y. S. H., and N. M. Jubeh. 2006. Comparison of performance of compressed-air energy-storage plant with compressed-air storage with humidification. *Proc IME J Power Energ* 220(6):581–588.

Oren, S. S., and S. Hong Wan. 1986. Optimal strategic petroleum reserve policies: A steady state analysis. *Manag Sci* 32(1):14–29.

Patel, M. R. 2006. *Wind and solar power systems.* Boca Raton, FL: CRC Press.

Pereira, M., and L. Pinto. 1991. Multi-stage stochastic optimisation applied to energy planning. *Math Program* 52(1-3): 359–375.

Prabhu, N. U. 1980. *Stochastic storage processes: Queues, insurance risk and dams.* New York: Springer-Verlag.

Read, E., and J. George. 1990. Dual dynamic programming for linear production/inventory systems. *J Comput Math* 19(11):29–42.

Rout, U. K., K. Akimoto, F. Sano, J. Oda, T. Homma, and T. Tomoda. 2008. Impact assessment of the increase in fossil fuel prices on the global energy system, with and without CO2 concentration stabilisation. *Energy Policy* 36(9):3477–3484.

Sanner, B., C. Karytsas, D. Mendrions, and S. Rybach. 2003. Current status of ground source heat pumps and underground thermal energy storage in Europe. *Geothermics* 32(4-6): 579–588.

Skogsberg, K., and B. Nordell. 2001. The Sundsvall hospital snow storage. *Cold Regions Sci Tech* 32(1):63–70.

Tang, L., J. Liu, A. Rong, and Z. Yang. 2001. A review of planning and scheduling systems and methods for integrated steel production. *Eur J Oper Res* 133(1):1–20.

Vinterback, J. 2004. Pell-Sim dynamic model for forecasting storage and distribution of pellets. *Biomass Bioenergy* 27:629–643.

Weiss, H. J. 1982. Economic Order Quantity models with nonlinear holding cost. *Eur J Oper Res* 9(1):56–60.

Winston, W. L. 1994. *Operations Research.* Boston: ITP Duxbury Press.

Yohanis, Y. G., and B. Norton. 2006. Including embodied energy considerations at the conceptual stage of building design. *Proc IME J Power Energ* 220(3):271–288.

Zanoni, S., and L. Zavanella. 2005. Model and analysis of integrated production-inventory system: The case of steel production. *Int J Prod Econ* 93-94(1):197–205.

5 Healthcare Supply Chain Management

Peter Kelle, Helmut Schneider, Sonja
Wiley-Patton, and John Woosley
Louisiana State University
Baton Rouge, Louisiana

CONTENTS

5.1   INTRODUCTION

A recent visit to a local emergency room (ER) motivated one of the coauthors to literally investigate the crucial element of supply chain management (SCM) within the healthcare system. After experiencing eight hours of delays in the ER waiting room and witnessing incredible lags in the medication dispensing process and even further impediments to the delivery of treatment modalities, the researcher became outraged with the thought of a high-performing healthcare system's lack of better supply chain management and service delivery. As a healthcare consumer, this scenario is likely to happen to all of us.

One might contemplate, how does an industry based on customer service fall short in the arena of inventory, patient, and materials management? According to business researcher Vicki Smith-Daniels (2006), in a field where precision is literally a matter of life and death, it seems strange that a crucial supportive function like inventory control and purchasing is often a hit-or-miss process. Healthcare, unlike other industries, has not given supply chain management the detailed attention that it so rightly deserves and needs to ensure patient safety and reduce overall healthcare costs.

Inventory management in the healthcare industry presents several interesting challenges both from a managerial and operational perspective. The stakeholder relationships, product considerations, and managerial and regulatory policies typically seen in healthcare are unique and worth investigation. Before delving into specific inventory control issues or demand characteristics that make healthcare SCM particularly difficult, it is important to recognize the magnitude of this industry.

5.1.1   MAGNITUDE OF HEALTHCARE INDUSTRY AND IMPACT
OF PHARMACEUTICAL COSTS

According to statistics published by the Centers for Medicare and Medicaid Services, healthcare spending topped $2 trillion or 16 percent of the gross domestic product (GDP) in the United States in 2005 (Centers for Medicare and Medicaid Services 2007). In addition, this percentage is projected to increase to 18.7 percent in ten years (Heffler et al. 2005; Catlin et al. 2007) state that healthcare expenditures increase at an annual rate of 6.9 percent. In addition, there has been a shift from public to private healthcare financing in the United States, as well as other countries around the world. Most of these efforts are an attempt to continuously find new ways of both controlling healthcare expenditures and paying for prescribed treatments and pharmaceuticals.

Especially given the current economic crisis, a growing amount of attention is being given to the rising costs of healthcare and specifically pharmaceuticals. Healthcare providers, insurance companies, government agencies, and consumers alike are forced to address this issue and to explore alternative methods of cost reduction or cost containment (Culyer 1990; Jönsson and Musgrove 1997; Marmor and Okma 1998). To gain a better understanding of the significance of this issue, it is prudent to first identify the magnitude of healthcare expenditures. According to the Plunkett Research Group (2008), "The health care market in the U.S. in 2007 was made up of hospital care (about $697.5 billion), physician and clinical services ($474.2 billion), prescription drugs ($229.5 billion), nursing home and home health ($190 billion), and other items totaling $668.8 billion."

As shown, the financial aspect of the healthcare and pharmaceutical industries is quite staggering. Much of the research in this field has focused on the relevant costs of healthcare (Organization for Economic Co-operation and Development [OECD] 1994, 1995, 1996; Comas-Herrera 1999; Castles 2004); however, Rothgang et al. (2005) examined data reported by the OECD from 1970 to 2002 to investigate trends in the level of governmental involvement with healthcare systems. Although some shifting has occurred over the years, convergence toward a mixed system of both public and private healthcare seems apparent.

A significant area of healthcare costs is the pharmaceutical area, which represented approximately 10 percent of annual healthcare expenditures in the United States and about $550 billion globally in 2007 (Plunkett Research Group 2008). Despite the size and importance of this industry around the world, especially in developed countries, the area of healthcare SCM and inventory management has been given relatively little attention. Several researchers have estimated that inventory investments in healthcare range between 10 percent and 18 percent of total revenues (Holmgren and Wentz 1982; Jarrett 1998). Any measures taken to control expenditures in this area can have substantial impacts on the overall efficiency of the organization and its supply network and, as a result, the profitability of healthcare providers.

In 2003, Guillén and Cabiedes examined the pharmaceutical policies of European Union (EU) countries from the mid-1980s through the 1990s. As explained in their research, the costs of pharmaceuticals continue to increase while countries struggle to find financing to support these drugs. They noted that much of the healthcare services are shaped by the pharmaceutical policies, and they also identified three reasons justifying a focus on pharmaceuticals when examining healthcare supply chains: the reliance of modern medicine on drugs to both prevent and cure sickness, the significance of pharmaceutical expenditures around the world, and the characteristics of the pharmaceutical industry (i.e., use of technology, innovation, etc.). Almarsdóttir and Traulsen (2005) also identified a number of reasons why pharmaceuticals deserve extraordinary consideration in controlling inventory. These specific characteristics make the pharmaceutical industry a very powerful force in its own right.

As with any industry, the range of products can vary tremendously depending on the market, customer demand, the scope of services, and other managerial decisions. For the purposes of this chapter and the specific research cases we present, our focus is mainly on the supply chain of pharmaceuticals.

5.1.2 Special Conditions and Terminology of Healthcare SCM

The stakeholders and their interrelationships, the product characteristics, and the policies employed all have some impact on healthcare SCM and inventory control. We explain some specific conditions and terminology used in healthcare in more detail in the following.

5.1.2.1 Stakeholders

Here we summarize some unique factors influencing how healthcare stakeholders interact, how they are controlled, and how they manage operations. A detailed analysis is provided in the next section.

Doctors are the primary caregivers in the healthcare system; however, it is important to recognize that physicians, in many cases, are contracted service providers. They are not actual employees of hospitals in which they work. Although this enables hospitals to expand their service capacities and offer better customer service, they must also relinquish some measure of control to these doctors. Autonomy, as it relates specifically to customer care, is valued above all else by doctors. Patients trust that physicians are prescribing treatments and medicines that will address their individual medical needs. Attempts to restrict the choices or influence the conditions of medical treatment are met with a great deal of resistance.

Another consideration is that, unlike many industries of this size, *hospital administrators* and pharmacy managers have to manage very complicated distribution networks and inventory control problems without the proper training or educational backgrounds to do so efficiently. This in no way implies that these individuals lack the intelligence to perform these tasks. Most hospital administrators and pharmacy directors are themselves doctors, which means they are highly skilled and educated in medicine; however, they are not the engineers or supply chain professionals who are commonly employed to manage similar systems in other industries.

The hospitals order the majority of their supplies from a selected *group purchasing organization (GPO)* that purchases directly from the production companies.

5.1.2.2 Products

Product formulary is the term used to identify the variety of drugs offered by the hospital pharmacy and is a source of both conflict and cost for the hospital. This formulary is comprised of specific medicines each designed to address a particular medical need, but physicians' opinions may vary on the most effective drugs for satisfying patient requirements. Providing prescription options for the doctors is important, both to the doctors and hospital; however, this increases the number of drugs that must be carried by the hospital pharmacy for patient treatment if it desires to maintain such a high level of customer service. In addition, the product formulary changes quite frequently as physicians' prescribing behavior reacts to advancements in medical research and technology.

It is also important to recognize that pharmaceuticals have a number of other requirements. They must be handled by trained personnel and experts in this field. Extraordinary resources are committed to developing these items, to manufacturing them, and to controlling their distribution and usage. All of these aspects are highly regulated by governments and other regulatory agencies, and they may require additional documentation. Finally, unlike other areas, customers have a limited role in the product selection process. As patients become more aware of alternative treatments, they may influence doctors as they prescribe medications; however, more times than not these choices are made for them.

Many of these items are also *perishable* and must be destroyed if not used. Perishable items have a limited shelf life and in many cases have special transportation and storage requirements. Part of the concern with pharmaceuticals is that outdated or expired items may be overlooked and dispensed to patients, which could have potentially disastrous effects both in patient care and public relations. The perishable item inventory control problem is a difficult one and has been studied many times using a periodic review approach (see Fries 1975; Nahmias 1975, 1982), with less attention being given to continuous review systems; however, there have been a number of works in this area in recent years (see Weiss 1980; Schmidt and Nahmias 1985; Chiu 1995; Liu and Lian 1999; Lian and Liu 2001). Although these studies provide valuable insight into the management of such products, none of these models have been tested in pharmaceutical inventory control. Given the combination of high costs and perishability of prescription drugs, it seems that more study is warranted as pharmacy managers look for help in setting optimal order policies.

5.1.2.3   Policies

Substantial efforts are made to regulate the healthcare industry. Government involvement is high, as it can provide both oversight of caregivers and funding for care recipients. Previously, healthcare systems have paid little attention to the management of inventories. However, with the implementation of *diagnosis-related groups (DRGs)* in 1983 by the United States government, these systems have turned their attention to cost containment as a means of increased profitability. The original objective of DRGs was to develop a patient classification system that related types of patients treated to the resources they consumed. Hospital cases are segmented into 1 of approximately 500 groups expected to have similar hospital resource use. DRGs are used to determine how much Medicare pays the hospital, since patients within each category are similar clinically and are expected to use the same level of hospital resources.

Interestingly, public and managerial policies targeted at cost control have failed in many cases to produce the necessary changes in stakeholder behaviors to achieve such outcomes. Efforts to control pharmaceutical prices have provided little relief as drug manufacturers attempt to recover the resources spent researching and developing these products. These companies also target physicians and patients with advertising campaigns as they attempt to influence prescribing behavior. Another reason policy has been somewhat ineffective at cost restraint is the conflicting stakeholder goals that appear throughout the supply chain. Doctors and hospital administration are often at odds as they try to balance the issues of prescribing autonomy and product variety. While the doctors want a large variety of brand-name drugs, the hospital management strives to minimize costs in the overall system by seeking generic drugs as substitutes for medicines preferred by physicians. Hospitals and GPOs have different objectives. Hospitals focus on negotiating the best prices for a wider selection of drugs from different pharmaceutical companies, while the GPOs are interested in larger orders from only a few pharmaceutical production companies.

Inventory control variables are called *par levels* in healthcare inventory management. The min par level is equivalent to the reorder point: if the inventory level decreases to or below this level, an order is triggered. The max par level is equivalent to the order up level (or base stock).

5.1.3   LIMITED SCOPE OF RESEARCH CONSIDERED

This chapter focuses specifically on SCM issues from the perspective of the hospital. We recognize the importance of other members of the supply network in this industry; however, the subject is simply too voluminous to cover adequately in this setting without restricting the scope somewhat. As such, the authors have selected to focus on supply chain problems as they are viewed by hospitals, which is appropriate given their significance in the supply chain and the costs they incur. In the following sections we identify critical managerial areas that distinguish the healthcare environment from other industries that have received attention in the past.

Pharmacy directors are turning their attention to controlling costs in the areas of SCM and pharmaceutical inventory control. Supply and purchased services account for the second largest cost component for a hospital, and it is clearly recognized that SCM is one of the principal areas for improvement in organizational performance. Supply chain management in hospital systems all over the world shows great variability both in performance and in focus. Such a variability combined with the growing cost of care and pharmaceuticals determines relevant differences in health service efficiency and unacceptable annual increases in healthcare costs.

The remainder of this chapter will serve to address several key areas of healthcare SCM and pharmaceutical inventory management. A review of the relevant managerial and quantitative modeling literature is provided. Here attention will be given to both the overall study of healthcare SCM and the quantitative approaches to inventory optimization. The practice of pharmacy operations will be described with specific attention given to the movement and distribution of prescribed medications within a hospital. Solutions focused on addressing SCM challenges will be presented and a specific case of a local hospital will be discussed in more details. Gaps in healthcare SCM and inventory research will be listed and recommendations for SCM research and best inventory control practices are suggested.

5.2   MANAGERIAL RESEARCH IN HEALTHCARE SCM

Traditionally researchers have taken two approaches to studying healthcare SCM: analyzing the managerial concerns and evaluating various quantitative models. First, the managerial aspects of such supply chain systems are discussed. The stakeholders in the healthcare supply chain and value chain are examined along with the conflicting goals in the system. Thereafter, the most common managerial approaches in healthcare SCM are summarized, including outsourcing and vendor managed inventory (VMI). In existing research, the focal point has primarily been on description of the system rather than developing innovative models or performing quantitative analyses. However, these descriptive studies are important to documenting the actual healthcare supply practices.

5.2.1 Stakeholders and Relationships

The financial aspects of the healthcare industry are significant, but it is also an industry very important to all members of society as it involves a wide range of stakeholders. Burns et al. (2002) examined the healthcare industry for three years in order to investigate its value chain, to uncover significant industry trends, and to identify the major stakeholder groups involved with healthcare services. Evidence showed that both vertical and horizontal integration were present in healthcare. Vertical integration was illustrated by hospitals teaming with insurance agencies or ambulatory services to combine various portions of this delivery network. Noticeable horizontal integration manifested in the form of hospitals purchasing other hospitals or in the formation of groups purchasing organizations. A significant result of this research was the identification of the following groups as stakeholders in this industry: (1) payers, (2) fiscal intermediaries, (3) providers, (4) purchasers, and (5) producers. Each of these groups is explained in more detail in the following.

Any person or organization responsible for supplying the funds to pay for medical expenses is identified as a payer. Based on this definition, examples of payers would include government, employers, individuals, and employer coalitions. Insurance agencies, health maintenance organizations (HMOs), and pharmacy benefit managers fall under the category of financial intermediaries. Any group supplying healthcare screening, treatment, or any other healthcare-related provisions are considered to be providers. Such entities would be physician offices, hospitals (or hospital systems), surgical centers, alternative and satellite facilities, ambulatory services, and pharmacies. Individuals or groups procuring any of these services are viewed as purchasers; however, it is important to note that this category extends beyond the basic consumers of healthcare services. This group also includes a wide variety of product resellers, independent distributors, pharmaceutical wholesalers, and so forth. The final group is the producers whose responsibility is the manufacturing of healthcare products, equipment, and technology. This includes pharmaceuticals, surgical equipment, information technology services, medical devices, and other capital equipment found throughout the healthcare system.

Tarabusi and Vickery (1998a, 1998b) stated that attitudes and habits of local pharmacists and physicians must be known for good access to markets. This research identified the importance of cost-containment programs in the pharmaceutical industry by comparing the approaches employed by the United States and countries of the European Union. Globalization and international partnerships have grown as pharmaceutical companies strive to control the research and development costs associated with these drugs.

5.2.2 VALUE CHAINS

The notion of value in the healthcare service value chain can be described by the interrelations of relevant stakeholders. Value chains were first introduced by Porter (1985) and are significant marketing tools because they afford managers the opportunity to assess the specific value added by each member of the supply network. Supply chains provide a mechanism for delivering that value to consumers. The original view of a value chain focused on a company's internal activities specifically designed to create value for customers; however, this concept has been extended to include the entire product and service delivery system (Bower and Garda 1985; Evans and Berman 2001).

Pitta and Laric (2004) provided a model of the healthcare value chain and supply chains, as they exist in many practical situations. Figure 5.1 illustrates that this supply chain is not linear or sequential in nature but closely follows the flow of information through the system. The success of the system or value created are linked to the transfer of quality information as the medical care received by patients relies heavily on information processing.

The first two groups interacting in this network are the patients and physicians. This stage of the process is generally initiated by the patient and provides valuable information necessary to adequately address whatever needs they might have. This research showed that individuals are much more likely to share very personal information with their healthcare providers when they believe this information is needed for medical purposes and when they trust the confidential nature of the patient–doctor relationship. The next link in the chain is created by the addition of pharmacists and other providers of medical equipment and services. In this stage the pharmacist creates value by further investigating the medical history of patients,

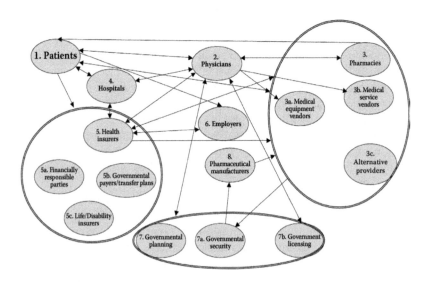

FIGURE 5.1 Stakeholders in the healthcare value chain (Pitta and Laric 2004).

specifically as it relates to medications, to determine any potential risks or interactions that may result from the addition of new, prescribed drugs. This is a very important step in customer service and patient care.

The next addition to the value chain is that of the hospital and the related services and procedures that may be included in diagnosing patient symptoms. As explained by Pitta and Laric (2004), doctors often request that batteries of tests be performed on patients even when symptoms and other indicators suggest a course of treatment. Hospitals create and store vast amounts of medical data for patients, which can be useful going forward. The fifth member of the healthcare value chain is the health insurers, which includes both public and private entities responsible for providing financial support to those receiving care. Since many insurers require a series of diagnostic tests and related data be completed before approving potentially costly procedures, patient data continues to grow. The addition of insuring groups or companies can also be a negative influence on the value chain. As one can see, with the addition of new members to the network and the creation of data at each stage, the healthcare system and participant interactions are becoming more and more complex.

Employers are the sixth group added to create another value chain. In the United States employers are the most often used source of medical insurance for employed persons and their families. Here the value is obtained by negotiating better benefits for groups of people as opposed to those offered to individuals. When the U.S. government introduced its federal Medicare and Medicaid programs in the 1960s, it became the largest medical insurance provider in the United States. A result of this involvement is the increasing influence of government regulation and policy in the healthcare industry. As such, government becomes the seventh participant of interest here as it influences various parts of this system. The final member of the model is the pharmaceutical manufacturers. It has already been established that pharmaceuticals represent a significant area of cost in the healthcare industry, as well as being the primary means of preventative and curative medical treatment.

5.2.3 STAKEHOLDER CONFLICTS

Physicians and pharmacists/pharmacy directors clash over medications offered by the hospital. The basic conflict here revolves around the issue of product variety versus economies of scale. The doctors have professional and personal preferences that are reflected in their prescription decisions. They value their individual freedom of choice in selecting the medications that they feel best address the specific needs of the patients under their care. Drug manufacturers, sales representatives, and the appearance of new drugs in the market also influence physicians. In contrast, the pharmacy directors pay more attention to the costs associated with specific medications and promote the usage of generic rather than brand drugs when available. They strive to take advantage of economies of scale whenever possible in the selection of drugs offered.

In 2005, Prosser and Walley examined the extent to which cost influenced prescribing behaviors of general practitioners (GPs) in the UK healthcare system. According to the authors, primary care organizations, (PCO) involvement has steadily increased as a means of providing more unified, controlled budgets for these physicians, as well as providing a monitoring mechanism of prescribing activities with this cost-containment objective in mind. The primary goal of this investigation was to measure the cost-awareness of doctors and to evaluate the attitudes of these GPs towards cost-restraint policies. Another goal was to determine what effect this managerial objective might have on their professional behavior. In this case the evidence demonstrated that all of the physicians were aware of pharmaceutical costs when prescribing treatments for their patients and felt that these costs should be considered when making prescription choices. However, there appeared to be a great deal of variation in the amount of influence these costs actually had on modifying medical decisions. As expected, physicians identified patient care as their primary focus during the treatment process with cost being a secondary concern.

Prosser and Walley (2007) continued their research in this area and employed qualitative methods to examine the managerial influence of healthcare administrators on the prescribing autonomy of general-care practitioners. In general, physician prescribing autonomy has been challenged recently due to the greater sophistication of patients in the ability to direct medical decisions and the increased bureaucratic involvement in service delivery in healthcare. This research showed that the objective of cost-containment presents a formidable obstacle for caregivers striving to offer the highest level of care and individualized service to those they serve.

This analysis showed that managerial involvement will have limited success in influencing GPs' prescribing autonomy. The conclusion was that the GPs directly resisted the attempts administrators made to modify their prescribing behaviors. The GPs maintain it is their responsibility to offer the best medications and treatment available to their patients, and that this autonomy is a component of providing individual care. In many cases this is an area of great conflict between doctors and hospital/pharmacy administrators, since larger product formularies desired by doctors usually result in higher inventory costs and reduce the opportunity for cost savings. When asked, doctors acknowledged the importance of controlling the costs of pharmaceuticals, but they identified the quality of patient care as being the paramount objective. Managers placed a great deal of importance on keeping prescribing costs within established budgets.

Another divergence appears between hospitals and the group purchasing organization (GPO) on the issue of product variety. Hospitals focus on negotiating the best prices for a wider selection of drugs. The attention is shown to brand-name medications in most cases. The GPO, on the other hand, strives to minimize costs in the overall system by seeking generic drugs as substitutes for medicines preferred by physicians. Again, a great deal of effort is given to the achievement of economies of scale for demanded items by considering a limited formulary (narrower selection of drugs).

The importance of product standardization for managing costs and improving clinical performance is outlined with respect to physicians' unwillingness to accept product alternatives to branded ones, even in the face of evidence regarding equivalence. As such, hospitals persist in the argument for wider drug selections. Both physician and nurse involvement and leadership in product analysis is necessary for successful standardization to occur, and for the provision of metrics pertaining to products and their relationship to the performance of the supply chain.

5.2.4 MANAGERIAL APPROACHES IN HEALTHCARE SCM

In this section of the chapter, we first explore the various strategic approaches that have been pursued in the area of SCM and inventory control. Some of the topics to be discussed are the outsourcing of distribution activities, allowing suppliers to manage inventory levels at various distribution points, and the use of common statistical techniques to achieve organizational and system goals.

5.2.4.1 Outsourcing

Kim (2005) presents an explanation of an integrated SCM system developed to specifically address issues related to pharmaceuticals in the healthcare sector. Many industries have recognized the importance of improved information sharing throughout the supply chain, and this work considers it as the most critical success factor. Here the supply network is composed of pharmaceutical companies, a wholesaler, and hospitals. The hospital's operating procedures and policies were reviewed to determine system requirements in an effort to improve the management efficiencies of the supply chain.

Jayaraman et al. (2000) present several tools and practical ideas to improve the flow of materials in a small healthcare facility. Traditional techniques, such as Pareto diagrams and department–product–type (DPT) matrices, were employed to track item flow and identify sources of errors or difficulties. Researchers suggested several procedural and policy changes be implemented to reduce inventory management problems. Landry and Beaulieu (2000) present a descriptive study of logistics systems at hospitals from three countries—France, the Netherlands, and the United States—to identify the best practices for replenishment policies, equipment, and handling technologies.

Nicholson et al. (2004) compared the inventory costs and service levels of an internally managed three-echelon distribution network and contrasted it with an outsourced two-echelon distribution network. Their research revealed that a general trend in healthcare is the outsourcing of specific organizational activities, inventory management, and materials distribution to "expert" third-party providers. To this point very little attention has been given to inventory management; however, changes in pharmaceutical regulations and the formation of the DRGs prompted a shift in approach.

When outside entities can provide needed products or services more efficiently than internal departments, it can prove very rewarding. In other instances, outside providers may demonstrate the ability to provide the desired products or services at a higher level of quality than an organization may be capable of achieving (Lunn 2000). A long-term benefit of outsourcing is the ability to reduce the number of suppliers in the system, which will eventually lower the procurement costs for the downstream members of the supply chain. As such, the short-term benefits of increased efficiency and higher quality, along with the long-term benefit of lower procurement costs, make outsourcing a logical approach to overall cost containment in the healthcare industry and consistent with current practices in inventory management (Veral and Rosen 2001).

Outsourcing decisions in this area are motivated by three factors: the magnitude of investment, the impact on service quality and delivery, and the availability of qualified service providers. Quality of service is arguably the paramount goal of healthcare providers (Li and Benton 1996), and it has been suggested that outsourcing various functions allows organizations to improve the quality of its internal operations. Jarrett (1998) identified several areas in which providers were able to improve internal performance. Patient care was improved, which can influence customer satisfaction and perceptions related to the quality of service provided by an organization. As the expertise develops and the capabilities of external sources grow, outsourcing becomes an increasingly attractive option. Another motivation for healthcare providers considering outsourcing the inventory management functions are the "success" stories similar to those described by Rivard-Royer et al. (2002).

Rivard-Royer et al. (2002) examined the changing role of distributors in the healthcare industry in recent years by investigating how operational processes had changed at a Quebec hospital As reported by Jarrett (1998), the ever-increasing desire to control healthcare costs have led hospitals to reevaluate many of their previous policies and operations. Inventory control is a logical area where substantial gains can be realized. Pharmaceuticals represent a significant cost equation related to both hospital inventory and quality patient care. As such, many hospitals and hospital systems have focused on this specific material management issue when restraining costs. U.S. pharmaceutical distributors were offering "stockless" replenishment where the distributors would prepackage items based on usage at individual care units (CUs), and would often provide direct delivery of drugs to these CUs (Henning 1980). By employing such a system, the distributors share in the benefits and risks associated with pharmaceutical savings or costs. The reduction in pharmaceutical inventory located at the hospital central pharmacy saved hospitals valuable resources, and the shifting of duties from the pharmacies to the distributors allowed the hospitals to reduce their workforce. Another improvement was noticed in enhanced customer service. However, as one would expect, a critical success factor in achieving any benefits from such a relationship and process is the continuous exchange of information between the point of use and supplier (Bolton and Gordon 1991).

5.2.4.2 Supply Coordination and Vendor Managed Inventory

According to Rivard-Royer and colleagues (2002), by the late 1990s stockless replenishment was a thing of the past as hospitals sought a balance between the amount of effort being spent in replenishing hospital inventories and the hospitals' inventory savings. The hospital involved with this study employed a hybrid approach since high-demand items were purchased in bulk (at the case level) for delivery directly to the patient CUs from distributors, and low-volume items were handled by the pharmacy central storage (CS). For these low-demand items, the CS was charged with breaking down the bulk purchases into smaller quantities for distribution to the various CUs as items were used. However, this hybrid approach was found to have limited benefits to both the hospital and distributors. Distributors were asked to perform additional work without receiving a comparable level of additional revenue. The hospital was still unpacking, repacking, and preparing drugs for distribution to the CUs and failed to realize significant workload reductions as a result of this replenishment approach. Overall, the gains were very limited in this particular study.

One managerial shift has been to vendor managed inventory (VMI), which has been shown to have benefits in SCM related to enhanced material handling efficiency. Other trends in supply coordination have included online procurement systems and the availability of real-time information sharing. These advances have demonstrated an ability to reduce total pharmaceutical inventory by more than 30 percent (Kim 2005). In addition, the improved information sharing throughout the supply chain allowed for more timely and accurate inventory data, which resulted in better demand forecasts and materials management.

The research of Meijboom and Obel (2007) investigated supply chain coordination in a global pharmaceutical organization. The authors looked at the issues facing a pharmaceutical company with a multilocation and multistage operations structure. Simulations are used to explore various coordination activities focusing specifically on "tactical control" in the firm. Their results suggest that a functionally organized operations structure will be unsuccessful in a multilocation, multistage environment. According to the authors, firms can employ one of two methods. They can either use a centralized approach relying heavily on information systems, or maintain a decentralized structure that relies on transfer prices as the coordination device.

Lapierre and Ruiz (2007) developed different modeling approaches, a cost minimization model and a balanced schedule model, for improving healthcare inventory management by examining the impact of scheduling decisions on the coordination of supply activities while recognizing inventory capacities. Specifically, the researchers concentrated on timing issues ranging from the purchasing and delivery of inventory to the work schedules and job assignments of employees. A significant outcome of this work is the promotion of coordinated activities throughout the purchasing and procurement process such that schedules can be expanded to include more specific details for improving the resource utilization across the organization. The models were tested in a hospital in Montreal, and hospital managers believed that the schedules generated by the models were both efficient and well balanced.

5.3 QUANTITATIVE MODELING IN HEALTHCARE SCM AND INVENTORY CONTROL

One of the primary goals of our initiative is to model and improve inventory management practices being employed by hospitals. There are a number of relevant inventory models in the current body of supply chain literature that are related to this effort, and those models are examined and described herein. In this review most of the attention will focus on models addressing multi-item, single-location, and multiechelon coordination given the unique conditions germane to the industry considered. Inventory models have been collected that have been formulated to address healthcare problems, have been applied, or have a high relevance to healthcare-related inventory management.

After summarizing the specifics of inventory control policies and planning frameworks in hospital pharmacies, the demand specifics are described in this section. A major portion of this section is devoted to inventory modeling; models of specific interest are multi-item–single-location and single-item–multilocation models. Literature associated with each of these modeling approaches is discussed, as well as how these models have been employed in the past.

5.3.1 INVENTORY CONTROL POLICIES AND PLANNING FRAMEWORK IN HEALTHCARE SCM

Efficiently managing pharmaceutical inventory systems requires another approach than the continuous review reorder point model. There are at least three limitations for using the continuous replenishment model in the context of healthcare supply systems: (1) the model does not account for the limited human resources; (2) it does not account for physical storage capacities, particularly the one at the CS level, which is critical in most hospitals; and (3) the decisions are only based on costs, not accounting for inventory control activities and their restricted capacity. Each of these limitations is elaborated on next.

Continuous replenishment inventory management is usually not applicable in hospitals since visits to CUs take a lot of the employees' time. To make it more efficient, tours are made such that several CUs are visited consecutively for rounds of stock control or supply distribution. The *periodic replenishment* model seems more appropriate for healthcare organizations, as each CU is replenished according to a schedule instead of when reaching a reorder point. However, with this inventory management approach, the frequencies of the visits are determined based on the order period and the economic order quantity (EOQ). Storage capacities at CUs and the CS are important factors when deciding how frequent a CU should be replenished and how frequent a supplier should be called. An undersized CS and open shelves able to contain very limited quantities of the voluminous products is often the reality most hospitals have to face. This situation calls for more frequent orders to suppliers and keeping more stocks at the CU than those suggested by the EOQ model in order to satisfy requirements and respect capacities. In that context, CU's stocks are kept higher to compensate for a small CS. Thus, greater volumes of items can be stored onsite by keeping these extra quantities. Another reason for keeping more stock in the overall system is that stockouts are time expensive for the supply service that needs to make extra emergency visits, internally referred to as hot-picks, to CUs, but also for the medical staff members who waste their time making extra calls to the supply department or chasing the products at other CUs.

One of the distinct features of materials management in a hospital is the use of a periodic review par level (or order-up to level) servicing approach. A major issue in setting par levels for various items in a healthcare setting is that these levels tend to reflect the desired inventory levels of the patient caregivers rather than the actual inventory levels needed in a department over a certain period (Prashant 1991). In most cases these par levels are experience-based and politically driven, rather than data driven. This poses a problem for warehouse managers, since the inventory they hold is typically based on aggregate hospital demands, while requirements of departments when aggregated are not in line with such estimates. The literature analyzing the setting of optimal par levels and review periods for multiple echelons draws upon prior work in multiechelon inventory systems, which are discussed in the next sections.

On the quantitative side, among one of the first healthcare inventory models was published by Michelon et al. (1994) who developed a tabu search algorithm for scheduling the distribution operations in a hospital. In their application, the number of replenishment visits is given a priori, and the problem consists of finding the schedule that minimizes the number of carriers given several time windows and practical constraints. Another investigation concentrating on this type of inventory control situation was performed by Banerjea-Brodeur et al. (1998) who looked at an application of a routing model to match the different CUs to be visited by a laundry department in a hospital. Here the authors used a combination of quantitative techniques and common sense to minimize the number of routes while respecting the volume capacity constraint of a cart.

Supply chain management can be fitted into a planning framework. Vissers et al. (2001) defined a planning framework for hospitals. One of the primary objectives of this research was to balance service and efficiency throughout the organization. As identified by the investigators, there are three characteristics that fit hospital production control settings. They are as follows: (1) a demand that is, generally speaking, larger than supply; (2) restrictions on supply defined by contracting organizations; and (3) higher patient expectations on service quality (Vissers et al. 2001). As with many organizations, the quandary then becomes how to maximize resources while still achieving higher levels of customer service. The developed framework consists of five levels of planning and control: patient planning and control, patient group planning and control, resources planning and control, patient volumes planning and control, and strategic planning (Vissers et al. 2001). The researchers describe the framework, as well as explain instances where current hospital policies and practices deviate from their suggestions and where changes may benefit the organization.

Another attempt is to include SCM into an enterprise resource planning (ERP) framework. Implementation of ERP in healthcare is very challenging due to the special characteristics to consider. Van Merode et al. (2004) explain the planning function of the hospital environment and identify facets of the healthcare industry that make ERP implementations and utilization more difficult. Some attention is given to explaining areas in which ERP can or cannot be used. According to the authors, hospital functions should be divided into "a part that is concerned only with deterministic processes and a part that is concerned with non-deterministic processes" (van Merode et al. 2004). It is their contention that the deterministic processes can be handled very well by ERP.

5.3.1.1 Demand Specifics for Hospital Pharmacy

The usage of pharmaceutical products in hospitals has specific characteristics. Typically, 40 percent to 60 percent are high-demand items that are relatively easy to handle based on usage statistics. However, much care is to be taken on the frequent changes in the medications used, and on the so-called formulary changes. New items come frequently, and an initial qualitative forecast is required. A large proportion of the items are only rarely used, and the usage times and quantities are frequently changing. Another group of items is used by patients for only a few days and finished afterward. Here the autocorrelation of the demand time series is challenging to consider.

Intermittent demand refers to the usage pattern of items that have extended time gaps separating irregular or sporadic periods of demand. If the quantities of the usage are also variable, we have the so-called lumpy demand that is a major challenge in pharmacy inventory management. The study of intermittent and lumpy demand has typically been approached from either a forecasting or inventory control perspective, with a number of studies incorporating both of these areas to examine practical issues related to inventory system performance. The body of work discussed in this section provides examples of this.

Sani and Kingsman (1997) provide a good review and compare various periodic inventory policies like the normal approximation (Roberts 1962), Naddor's heuristic (Naddor 1980), power approximation (Ehrhardt 1979), and several others. In addition, a variety of forecasting methods were reviewed with the authors attempting to determine which are best for low- and intermittent-demand items. Before discussing specific inventory control approaches or models, it is important to recognize two fundamental notations commonly used in this literature. The minimum inventory parameter, or the reorder point, is commonly denoted as s, while the maximum inventory parameter, or the order-up-to value, is represented by S. In situations where item demand appears to be low or intermittent, variations of the (s, S) form of the periodic review inventory control system have been promoted as the optimal approach to inventory management. As the authors state in their review, forecasting demand for such items can also prove very challenging.

The research presented by Sani and Kingsman (1997) examines the management of spare parts for agricultural machinery, which includes many items with seasonal demand characteristics displaying vastly different demand from summer to winter months. After examining the various (s, S) inventory policies with respect to annual inventory costs and customer service levels, Ehrhardt's (1979) power approximation proved to be a good inventory system for managing items with low overall demand. This method not only scored well in evaluations of cost minimization but also offered reasonable service levels for items with low to medium demand. From the comparisons, it seems that the Croston estimator (Croston 1972) is one simple forecasting method that performs adequately in most cases (using the exponentially smoothed interdemand interval, updated only if demand occurs). One surprising result was that a basic twelve-month moving average of demand yielded adequate forecasting results when examining both annual costs and service level, with exponential smoothing performing the worst on this measure; however, exponential smoothing performed best on the customer service level measure alone. It is important to note that both of the aforementioned techniques were utilized: updates every review period and the Croston forecast. Finally, the authors concluded that achieving customer service levels greater that 95 percent for lumpy demand items is impossible unless high materials stocks are kept as a means to satisfy this objective.

Hollier et al. (2005) developed a modified (s, S) inventory model to address cost control issues specific to lumpy demand patterns. This approach integrated a maximum issue quantity restriction and a critical inventory position as constraints influencing the inventory control policy. Here the primary objective was to minimize the system replenishment costs. These authors applied two algorithms, one being a tree search and the other a genetic-based algorithm to optimize the decision variables. The numerical examples and results illustrate the benefits of employing such algorithms, as well as demonstrate the utility of the maximum issue quantity and critical inventory position constraints when managing lumpy demand items.

Syntetos and Boylan (2006) employ simulation models to provide an evaluation of various forecasting techniques, specifically simple moving average, single exponential smoothing, Croston's estimator, and a new technique introduced in their paper, when handling lumpy or intermittent demand items. Service level and stock volume were evaluated using a number of performance measures related to customer service and inventory costs. In this case the authors reported that the new estimator they developed outperformed other forecasting techniques as an inventory control method; however, it is important to recognize that the simple moving average technique yielded favorable results as well compared to the other two methods.

5.3.1.2 Multi-Item, Single-Location Models for Pharmacy Inventories

We found few papers that applied multi-item models to control healthcare inventories. Dellaert and van de Poel (1996) derived a simple inventory rule, a (R, s, c, S) model, for helping buyers at a university hospital in the Netherlands. The notations of s and S were as previously defined. Here R represents the length of the review period (time between orders) and c is the can-order level. Since most items have a joint supplier and the orders for a certain supplier are always placed on the same day of the week, they extended an EOQ model to a so-called (R, s, c, S) model, in which the values of the control parameters s, c, and S are determined in a simplistic manner. This approach resulted in substantial gains, which were observed in improved service levels, reductions in supplier orders, smaller total inventory levels and holding costs, and substantially lower system costs, for the participating hospital.

However, we could not find other multi-item inventory applications in healthcare. We summarize next the models that seem to have the best application potential for this specific area. The majority of results for the multi-item problem are for the deterministic demand case where the sizes and the multiproduct sequence of orders have to be determined. The problem is known to be NP-complete (nondeterministic polynomial time complete) and therefore many heuristics have been proposed. For an overview on the approaches, see Gallego et al. (1996), Hariga and Jackson (1996), and Minner and Silver (2005).

The early investigations in stochastic multiproduct inventory problems were generated by Veinott (1965), Ignall (1966), and Ignall and Veinott (1969). Later, in the study of Beyer et al. (2001), the researchers examined the management of multiple items with a warehousing (storage) constraint, which was based on the early model of Ignall and Veinott (1969), and generated results for finite-horizon and infinite-horizon discounted-cost problems. This work demonstrates optimal policies suitable for the various conditions that occur in healthcare.

Ohno and Ishigaki (2001) examined a continuous review inventory system for multiple items exhibiting compound Poisson demands and created a new algorithm for determining the optimal control policy. This alternative method was derived using the policy iteration method (PIM) and resulted in a substantial decrease in processing time needed to evaluate and improve the optimal policy. In addition, the new algorithm was tested using three joint ordering policies.

Minner and Silver (2005) analyzed the stochastic demand, continuous review lot-size coordination problem for Poisson demand, and negligible replenishment lead times. A formulation as a semi-Markov decision problem was presented to find the optimal replenishment policy and several heuristics were suggested and tested in a numerical study. However, the results only apply for pure Poisson demand and cycle inventories (i.e., without safety stocks). With the assumption of negligible lead times, safety stocks here serve to avoid the negative consequences of transaction sizes that exceed available inventory rather than covering against demand uncertainty over the replenishment lead time.

In a recent paper, Minner and Silver (2007) analyzed a replenishment decision problem where each replenishment has an associated setup cost and inventories are subject to holding costs. The solution of this trade-off results in order batch sizes. The warehouse space for keeping inventories is limited, which generally restricts these batch sizes. A second aspect of their analysis is that demands are random, here being modeled by a compound-Poisson demand process. This further complicates the analysis because now safety stocks and cycle stocks share the limited warehouse space.

5.3.3.3   Single-Item, Multilocation Models: Multiechelon Coordination in Pharmacy

The literature analyzing the setting of optimal par levels and review periods for multiple echelons draws upon prior work in multiechelon inventory systems, which are discussed next.

One of the first investigations in the area of multiechelon distribution networks was performed by Allen (1958), who attempted to optimally redistribute stock among several locations. Since then, numerous works extended this model (see Simpson 1959; Krishnan and Rao 1965; Das 1975; Hoadley and Heyman 1977). Another significant effort in this area was conducted by Clark and Scarf (1963), as it was the first study that attempted to generate and depict an optimal inventory policy in a multiperiod, multiechelon, distribution model involving uncertain demand. Other research in this area has focused on the setting of optimal lot sizes and inventory safety stocks in a multiechelon supply chain (Deuermeyer and Schwarz 1981; Eppen and Schrage 1981; Nahmias and Smith 1994).

Sinha and Matta (1991) and Rogers and Tsubakitani (1991) present modeling studies in this research domain. Both studies focused on two-echelon inventory systems employing periodic review under stochastic demand with fixed lead times. Rogers and Tsubakitani concentrated on finding the optimal par values for the lower echelon such that the penalty costs were minimized. A budget value acted as a constraint of the maximum inventory investment. The optimal par level is achieved by utilizing a critical ratio that is adjusted by the Lagrange multiplier subject to the budget constraint. Sinha and Matta focused on minimizing the holding costs of multiple products at both echelon levels with the presence of penalty costs at the lower echelon. Similarly, the results demonstrated that the optimal par values for the lower echelon items were obtained using a critical ratio; however, the holding cost function provided a method for generating the optimal values for the upper echelon.

Nicholson et al. (2004) extend the work of Rogers and Tsubakitani (1991) and Sinha and Matta (1991) by considering a three-echelon inventory system. Specifically, this research concentrated on the healthcare sector and sought to compare an internally managed three-echelon system to an outsourced two-echelon distribution system. Comparisons focused on inventory costs and service levels. The results suggest that outsourcing the targeted functions yielded lower inventory costs without sacrificing customer service. As mentioned previously, the use of periodic review par level servicing at the departmental level (i.e., at the individual care units) is a unique characteristic to hospital material management; however, it is important to note that the central pharmacy often operates in a similar manner utilizing another set of par values reflective of aggregate inventory requirements. Zhu et al. (2005) found similar results in their study by using numerical simulations and sensitivity analysis of two models. Like that of Nicholson et al. (2004), this work demonstrated cost savings with high service levels for the two-echelon distribution network where item demands were monitored by the distributor and delivered directly to the departments for use, as opposed to orders being routed through the CS as they would be in a three-echelon system.

Some models consider the case of lost sales. Nahmias and Smith (1994) examine a two-echelon supply chain where stockouts can penalize the retailer in cases where customers are unwilling to wait for their orders to be filled. The authors assume instantaneous deliveries from the warehouse to the retailer, which is often not the case in practice. Andersson and Melchiors (2001) executed a similar study that allowed for excess demand to be handled using back orders. The supply chain in this work was comprised of one warehouse with multiple retail outlets being serviced. Demand here is independent Poisson processes, and lead times are considered constant. The heuristic developed by the authors provides a mechanism to evaluate this type of supply network on elements of cost and service level.

Another research study examining the lost sales case is that of Hill et al. (2007), which looked at inventory control in a single-item, two-echelon system with a continuous review policy. Again, a central warehouse services several independent retailers and then has its stock replenished by an external supplier. The system operates such that any excess customer demand at the retailer can be filled from the warehouse provided the item is in stock; however, any items that are not in stock result in lost sales. Lead time for the retailer is equal to the transportation time required to move the item from the warehouse to the specific retail outlet.

The influence of order risk was examined by Seo et al. (2001) in a two-echelon distribution system. They claim that the service policies should reflect the availability of real-time stock information. Their model adjusts the reorder time on the basis of an approximated order risk, which is associated with orders that are filled immediately versus those that are delayed. Results show that this order risk policy performed well when warehouse lead time was short, where item demand was low, and where there was an intermediate number of retailers in the supply chain.

Axsäter (2003) used an approximation technique to optimize inventory control in a two-echelon distribution network. Items displayed stochastic demand, and the system relied on a continuous review (R, Q) policy. In this situation, holding costs at all locations and back-order costs at the retailer were assumed to be linear. Axsäter presents a simplified method for estimating reorder points by using normal approximations for demand at both the retailer and the warehouse. Lee and Wu (2006) examined a simplified two-echelon supply chain system comprised of one supplier and one retailer. Here the retailer has a choice between two restocking policies: traditional methods (such as an EOQ or periodic review approach) or a statistical process control (SPC)–based method. The results show support for the SPC method in addressing inventory variation issues and in reducing order backlogs. Benefits are apparent in the areas of demand management and inventory control, which leads the authors to suggest this as an alternative method for managing supply chain and inventory costs.

5.4   PRACTICAL SOLUTIONS FOR INVENTORY MANAGEMENT OF PHARMACEUTICALS IN HOSPITALS

The following section discusses some practical solutions to issues facing managers when dealing with prescription drugs. The manner in which hospital pharmacies and drug distribution/dispensation historically occurred is examined. First we discuss the traditional method and then the current trends. We also describe the dilemmas hospital personnel face in filling prescribed medications and the alternative methods employed to satisfy drug orders. The managerial predicaments associated with managing skilled labor in this system are also explained in limited detail. We close the section with the summary of the main results of a project targeted to provide decision support to inventory management of a hospital pharmacy.

5.4.1   TRADITIONAL OPERATIONS IN HOSPITAL PHARMACIES

Traditionally, hospital pharmacies operated in a centralized manner. Medications were stored in a central pharmacy and then distributed by pharmacy technicians using dose carts to make deliveries to the various hospital floors and CUs. A setup such as this requires large amounts of inventory to be stored in a single location while also allowing for quick, easy access to items. These technicians were charged with the task of moving the various medicines in the correct amounts or in individual cassettes for each patient. These dose carts contained individual cassettes of medications for each patient, which the nurse or caregiver was then charged with administering to the patient. If for some reason the patient's drug order changed, a special trip would be required to retrieve the necessary medications and then return to the CU. Thus, changes in medication or dosage may have taken hours to satisfy, resulting in time delays in care, increased labor requirements, and an overall increase in system costs associated with wasteful or excess operations.

To avoid prolonged delays in administering medications, it was common practice for nurses to borrow a medication from another patient. That resulted in additional problems in the distribution process and potentially dangerous situations where patients may have received the improper medications. Other difficulties associated with the specific pharmaceutical types existed. Specifically, federal and state regulations require a variety of documentation associated with the dispensation of narcotic drugs. From the time a narcotic is delivered to the hospital until it is administered to the patient, the drug must be tracked by the hospital and is surrounded by a number of tasks such as checking the ordered medication against the patient's medical record, searching for narcotic keys to the cabinet, documenting for administrative records, and reconciling narcotic records after each shift. This example demonstrates some specific challenges in controlling and monitoring items in healthcare SCM.

Pharmacists are highly paid professionals being asked to spend long amounts of time handling excessive documentation and a labor-intensive drug dispensation and distribution process rather than practicing pharmaceutical care. The opportunity to reduce repetitive and tiresome tasks was very appealing to administrators and pharmacists alike. As such, the ability to automate these processes and take advantage of computer information systems was valued by stakeholders.

5.4.2 Current Management Trends and Operations in Pharmacy SCM

Hospitals have typically employed a variety of methods and policies to resolve pharmaceutical inventory management issues. A few of the more prominent solutions are discussed in this section to demonstrate the modern approaches seen in industry. Specifically, outsourcing, VMI, and information technology (IT)–based solutions are presented.

5.4.2.1 Outsourcing

As described in the previous literature review, outsourcing has been widely used in the healthcare industry. Regardless of the specific product, entering into partnerships with suppliers and distributors for the purposes of combining services have generated benefits for the healthcare providers (see, e.g., Li and Benton 1996; Jarrett 1998; Lunn 2000; Veral and Rosen 2001; Nicholson et al. 2004). The magnitude of resources linked to pharmaceutical inventory and its management, the desire to shift or reduce pharmacy workloads, and the opportunity to refocus resources on patient care make this an appealing proposition for healthcare providers. However, as shown by Rivard-Royer et al. (2002), all parties must benefit from such arrangements to provide long-term gains.

5.4.2.2 VMI

Another trend is the growing usage of VMI strategies (Kim 2005). This a specific type of outsourcing in which pharmaceutical inventories located at various distribution locations (i.e., at the CUs) around the hospital are monitored by the supplier, in this case the pharmaceutical company or distributor, and replenished as needed. Currently, several hospitals employ a continuous review (s, S) inventory control policy. When demand for an item reaches a predetermined minimum level (s), an order is automatically generated and transmitted directly to the supplier. The supplier, in turn, ships the amount necessary to refill the distribution centers to the maximum quantity (S). Depending on the specific circumstances, materials can be either sent to the pharmacy for repacking and distribution or sent directly to the point of service, which bypasses the pharmacy entirely.

5.4.2.3   IT-Based Inventory Management Solutions

Information systems play a significant role in all of the aforementioned suggested approaches; however, they are perhaps even more critical in the next solution. Perini and Vermeulen (1994) reviewed a number of devices (e.g., Lionville CDModule, Meditrol, Argus, MedStation™, Sure-Med, and Selectrac-Rx) focused on the dispensing of medicines located around the hospital in the patient care unit with the purpose of replacing the traditional dosage carts used by pharmacies, and to shift control of locally stored pharmaceutical inventories and controlled substances to caregivers at the point of use. Inventory stored in the CUs offer caregivers the opportunity to dispense medications quickly to patients; however, restocking these units can take extra time. These medication-management machines are designed to offer financial and practical advantages over traditional operating procedures where inventories are stored in a central pharmacy and then distributed by dose carts as needed. Another benefit of using these local devices is the ability to quickly create, store, and access point-of-service patient information, which can also expedite the documentation requirements associated with drugs. The technology employed by these solutions enhances operating efficiency and facilitates customer care by lowering the risk of patients receiving incorrect medications. Regardless of the specific solution, these systems usually offer pharmaceutical administrators the ability to reduce inventory carrying and other costs, to improve billing and usage information, and to increase staff productivity by creating a highly integrated, data-driven information flow.

5.4.3   A CASE OF HOSPITAL PHARMACEUTICAL INVENTORY MANAGEMENT

In this section we summarize the main results of a project targeted to provide decision support to inventory management of a hospital pharmacy. The available equipment and current management system are described first; the new quantitative models developed by the authors of this chapter are summarized next; and the managerial decision support potentials close the section (see the details in Kelle et al. 2008).

5.4.3.1   Pharmaceutical Distribution Equipment and Control Case

The drugs are stored in the local depots (Pyxis© MedStations) in close to a hundred different areas of the hospital. Each area has a different selection of drugs. The Pyxis MedStation registers each transaction date, the quantity of demand (withdrawal), each delivery (refill), and the actual inventory level. It is connected to the central depot (Tallyst© system) electronically.

As the inventory level decreases to a reorder point (min par level) an automatic order is triggered by the local depot. The order quantity is determined by the order-up-to level (max par level). The par levels can be selected and fixed for each drug. Currently, the par levels are based on so-called experiences, but no modeling or optimization is involved in setting these control values. The central depot satisfies the order within a day. Shortage in the central depot is rare, and in that case the wholesaler (GPO) refills within a day or an emergency supply is provided.

The demand for each drug is uncertain. The daily usage data is available for each drug for the period of one year. High service level is essential in the drug supply. In case of a shortage at a local depot (Pyxis), a very costly emergency delivery is necessary. The refill from the central depot is also constrained and expensive because a highly educated pharmacist must supervise it. One major goal of the pharmacy management is to minimize the number of refills per day. If the refill per day is very large, it cannot be done in two shifts, and to provide an extra shift is difficult and costly.

The capacity (the total space) of the Pyxis systems is limited. It consists of drawers. Each drawer is subdivided by spacers that can be relocated so different cubicles are constructed and assigned to each drug. Different drugs cannot be stored in the same cubicle. The total room for the cubicles is fixed. The inventory holding cost factor is not a major concern for the hospital because the wholesaler is financing it. However, we will include inventory holding cost later and will examine the effect of its consideration.

The main goals of the management are to:

- Provide a high service level for each drug
- Minimize the total number of expected refills (orders) per day for a Pyxis
- Use the limited space of a local depot (Pyxis) in the best way by subdividing it to separate areas (cubicles) for each drug

The decision variables are the reorder point (the min par level) and the order up level (max par level) for each drug. The average daily demand and the standard deviation of the daily demand can be calculated from historical data. Also, we know the storage space requirement for a unit of each drug and the maximum storage space in a Pyxis.

5.4.3.2 Quantitative Modeling Issues

We formulated different optimization models. In the first one we minimize the total number of expected refills (orders) per day for a Pyxis MedStation subject to two constraints:

- The service level for each drug is a fixed high level
- The total space needed for the maximum inventory level (Si) for each drug is not more than the available total space of the Pyxis MedStation

We have some challenges that are not handled in literature including the large dimensions (around 250 drugs in a Pyxis) and demand uncertainty; the service-level constraint combined with space constraint; and the fast solution and simple sensitivity analysis requirement. For this reason we provided a simplified model and solution for the optimization problem. We consider the two constraints (service level and space) and the two sets of decision variables separately, and solve it iteratively using power approximation (see Ehrhardt 1979) for handling the service level.

In the first step we set the reorder point, based on required service level, for each item of a Pyxis station, using power approximation formulae. The achieved service level depends also on the reorder level, which depends on the cost factor. We can show that the optimal order quantity implicitly (by the analogy of the EOQ model) sets the optimal cost rate. We need to apply an iterative solution starting with an initial estimate. We can select the initial estimate for the reorder point without safety stock consideration as the expected lead-time demand and provide the initial estimate. In the second step we minimize the number of orders per day under the space constraint considering the reorder points set before (see the details in Kelle et al. 2008).

5.4.3.2 Managerial Decision Support

The result shows that the optimal allocation of the space for the order quantities of the items is proportional with the square root of the demand over space rate. It is proportional to the available free space (space remaining for cycle stock, after taking out the space requirement for the reorder point stock). The reorder point stock is the sum of expected demand during the lead time plus the safety stock. The lead time is the refill time.

The trade-off between the number of orders and free storage space for a given service level can be expressed in a simple functional form depending on the remaining free storage space for cycle stock (above safety stock), so our function implicitly includes the service level trade-off. For a fixed service level, the trade-off curve has a hyperbola shape that is shifted to the right (higher space requirement) with increasing service level requirement. This property allows a simple trade-off analysis including managerial questions such as:

- Service level versus expected number of daily orders (worker capacity requirement)
- Number of drugs stored (formulary) versus service level
- Storage volume versus service level and worker capacity requirement
- Extension of formulary versus additional worker capacity requirement

5.5 SUMMARY AND RESEARCH PROSPECTIVE

This chapter was introduced with a scenario of an ER supply problem which, with variations in details, will possibly be experienced at some time by individuals who read this book. The researchers have presented a broad view of characteristics that make healthcare supply chain management a very intriguing area of study. The healthcare industry operates in a unique manner in that it has a number of special constraints all being enforced at once. Specifically, the focus is on maintaining a high level of service and controlling costs, as opposed to reducing them. In traditional supply chains, these variables are manipulated to achieve the optimal economic operating setting. However, in the healthcare industry, trade-offs are not as easily achieved. Furthermore, medicines are special products as noted by Almarsdóttir and Traulsen (2005).

First, the significance of the healthcare industry is demonstrated by the provided cost breakdown and by conveying the importance of affordable, quality care to the described stakeholders. Second, a review of current literature in this specific area is presented. It is important to recognize that the need for continued research in this domain is warranted. Many of the proposed inventory control models have yet to be tested in a healthcare setting or have not been evaluated under conditions of varying demand types (i.e., intermittent or autocorrelated demand). Finally, traditional operations of hospital pharmacies are described and contrasted with some common managerial approaches seen in practice today. Specifically, outsourcing and VMI in some form are used extensively to alleviate strains on pharmacy resources and material management issues. Information technology plays a significant role in determining the success of these solutions, but the widespread use of materials-management machinery in hospitals and throughout hospital systems makes this task even more daunting. It seems clear that investments in these resources will continue in the future as hospitals increase their reliance on the information created and controlled by these machines.

The body of work in healthcare SCM provides a good foundation for further study in this area; however, there are a number of gaps that should be addressed. The managerial literature in this area has generally focused on identifying stakeholders and relationships within the healthcare system and distribution network, with other works exploring trends in funding mechanisms and the influence of cost-containment policies on this industry. Very few best practices and benchmarks have been identified for healthcare managers to use as performance guidelines or indicators. We believe that future research targeting the changing relationships of stakeholders and supply chain partners should be pursued to offer valuable insight into the impacts of outsourcing and other supply chain coordination techniques currently in use. Both outsourcing and VMI are used extensively in modern healthcare supply chains, but little is known about the influence these business relationships have had on organizations or their operations outside of cost. As many hospitals employ outsourcing and IT-based medicine distribution machines as a means of simplifying their inventory management systems, they are actually adding to the complexity of their pharmaceutical distribution network. To this point, little is known about these arrangements or their implications.

In addition to the managerial literature, there are a number of gaps in the quantitative modeling research as well. Although there are a number of inventory control models that may be applicable to pharmaceutical management within hospitals, many of these models have yet to be tested in this setting. The special product characteristics mentioned in the earlier part of the chapter present unique challenges to managing such inventories and warrant further study. In addition, the use of the medicine machines at local care units (like Pyxis machines) suggests that multiple constraints must be considered when determining the control values employed at both the local and central storage areas. Service level, not inventory holding costs, is the primary concern in this system. Models combining space, workload, cost, and service-level constraints have yet to be developed; however, efforts in this area could prove invaluable to pharmaceutical managers as they attempt to focus on supply chain operations that offer cost-containment benefits. It is also important to develop models showing the trade-off between the extension of the formulary (variety of drugs provided) versus additional cost, workload, storage capacity requirement, and service problems.

To that end, the authors of this chapter are currently examining such problems and developing models to facilitate multicriteria decision support.

In conclusion, as we strive to comprehend and mend healthcare's woes we must remember that healthcare spending in the United States reached $2.3 trillion in 2007 and is projected to reach $3 trillion in 2011. Healthcare spending is projected to reach $4.2 trillion by 2016 (Poisal et al. 2007). The United States spends six times more per capita on the administration of the healthcare system than its peer Western European nations (McKinsey Global Institute 2007). The *McKinsey Quarterly* (2004) reported that health insurance expenses are the fastest growing cost component for employers. Our endeavor while academic in nature cannot heal all of healthcare's illnesses; however, our goal is to provide a practical model for executive decision support specifically in pharmaceutical inventory control. Our contention is that this research model will help to optimize inventory levels, maximize space usage, reduce pharmacy personnel workloads, and maintain high customer-service levels, thus decreasing cost and enhancing efficiencies in healthcare.

REFERENCES

Allen, S. 1958. Redistribution of total stock over several user locations. *Nav Res Logist Q* 5(1):51–59.

Almarsdóttir, A., and J. Traulsen. 2005. Cost-containment as part of pharmaceutical policy. *Pharm World Sci* 27(3):144–148.

Andersson, J., and P. Melchiors. 2001. A two-echelon inventory model with lost sales. *Int J Prod Econ* 69(3):307–315.

Axsäter, S. 2003. Approximate optimization of a two-level distribution inventory system. *Int J Prod Econ* 81-82:545–553.

Banerjea-Brodeur, M., J.-F. Cordeau, G. Laporte, and A. Lasry. 1998. Scheduling linen deliveries in a large hospital. *J Oper Res Soc* 49(8):777–780.

Beyer, D., S. P. Sethiand, and R. Sridhar. 2001. Stochastic multiproduct inventory models with limited storage. *J Optim Theor Appl* 111(3):553–588.

Bolton, C., and J. Gordon. 1991. *Health care material management*. Working paper, Queen's University, Kingston.

Bower, M., and R. A. Garda. 1985. The role of marketing in management. *McKinsey Quarterly*, Autumn 85(3):34–46.

Burns, L., and Wharton School Colleagues. 2002. *The health care value chain producers, purchasers, and providers*. San Francisco: Jossey-Bass.

Castles, F. 2004. *The future of the welfare state: Crisis myths and crisis realities*. Oxford: Oxford University Press.

Catlin, A., C. Cowan, S. Heffler, B. Washington, and the National Health Expenditure Accounts Team. 2007. National health spending in 2005: The slowdown continues. *Health Aff* 26(1):142–153.

Centers for Medicare and Medicaid Services, Office of the Actuary, National Health Statistics Group. 2007. 2005 National Health Care Expenditures Data. http://www.cms.hhs.gov/nationalhealthexpenddata/01_overview.asp? (accessed October 20, 2007).

Clark, A., and H. Scarf. 1963. Optimal policies for a multi-echelon inventory problem. *Manag Sci* 6(4):475–490.

Chiu, H. 1995. An approximation to the continuous review inventory model with perishable items and lead times. *Eur J Oper Res* 87(1):95–108.

Comas-Herrera, A. 1999. Is there convergence in the health expenditures of the EU Member States? In *Health care and cost containment in the European Union*, edited by E. Mossialos and J. Grand, 197–218. Aldershot, UK: Ashgate.

Croston, J. 1972. Forecasting and stock control for intermittent demands. *Oper Res Q* 23(3):289–304.

Culyer, A. 1990. Cost containment in Europe. *In Health care systems in transition: The search for efficiency*, edited by OECD, 29–40. Paris: OECD.

Das, C. 1975. Supply and redistribution rules for two-location inventory systems: One period analysis. *Manag Sci* 21:765–776.

Dellaert, N., and E. van de Poel. 1996. Global inventory control in an academic hospital. *Int J Prod Econ* 46-47:277–284

Deuermeyer, B., and L. Schwarz. 1981. A model for the analysis of system service level in a warehouse-retailer distribution system. In *Multi-level production/inventory control systems: Theory and practice*, edited by L. Schwarz, 163–193. Amsterdam: North-Holland.

Ehrhardt, R. 1979. Power approximation for computing (s, S) inventory policies. *Manag Sci* 25(8):777–786.

Eppen, G., and L. Schrage. 1981. Centralized ordering policies in a multi-warehouse system with lead times and random demand. In *Multi-level production/inventory control systems: Theory and practice*, edited by L. Schwarz, 51–67. Amsterdam: North-Holland.

Evans, J., and B. Berman. 2001. Conceptualizing and operationalizing the business-to-business value chain. *Ind Market Manag* 30(2):135–48.

Fries, B. 1975. Optimal order policy for a perishable commodity with fixed lifetime. *Oper Res* 23(1):46–61.

Gallego, G., M. Queyranne, and D. Simchi-Levi. 1996. Single-resource multi-item inventory systems. *Oper Res* 44(4):580–595.

Guillén, A., and L. Cabiedes. 2003. Reforming pharmaceutical policies in the European Union: A "penguin effect"? *Int J Health Serv* 33(1):1–28.

Hariga, M., and P. Jackson. 1996. The warehouse scheduling problem: Formulation and algorithms. *IIE Trans* 28(2):115–128.

Heffler, S., S. Smith, S. Keehan, C. Borger, M. Clemens, and C. Truffer. 2005. U.S. health spending projections for 2004–2014. *Health Affairs*. http://content.healthaffairs.org/cgi/content/abstract/hlthaff.w5.74#otherarticles (accessed on October 14, 2007).

Henning, W. 1980. Utilizing suppliers to the hospital's best interest. *Hosp Mater Manag Q* 1(3):39–47.

Hill, R., M. Seifbarghy, and D. Smith. 2007. A two-echelon inventory model with lost sales. *Eur J Oper Res* 181(2):753–766.

Hoadley, B., and D. Heyman. 1977. A two-echelon inventory model with purchases, depositions, shipments, returns, and transshipments. *Nav Res Logist Q* 24(1):1–20.

Hollier, R., K. Mak, and K. Yiu. 2005. Optimal inventory control of lumpy demand items using (s, S) policies with a maximum issue quantity restriction and opportunistic replenishments. *Int J Prod Res* 43(23):4929–4944.

Holmgren, H., and J. Wentz. 1982. *Material management and purchasing for the health care facility*. Washington, DC: AUPHA Press.

Ignall, E. 1966. Optimal policies for two-product inventory systems, with and without setup costs. PhD diss., Cornell University.

Ignall, E., and A. Veinott. 1969. Optimality of myopic inventory policies for several substitute products. *Manag Sci* 15(5):284–304.

Jarrett, P. 1998. Logistics in the health care industry. *Int J Phys Distrib Logist Manag* 28(9):741–742.

Jayaraman, V., C. Burnett, and D. Frank. 2000. Separating inventory flows in the materials management department of Hancock Medical Center. *Interfaces* 30(4):56–64.

Jönsson, B., and P. Musgrove. 1997. Government financing of health care. In *Innovations in health care financing*, edited by G. J. Schieber, 41–64. Washington, DC: World Bank.

Kelle, P., H. Schneider, S. Wiley-Patton, and J. Woosley. 2008. Pharmaceutical supply chain specifics and inventory solutions for a hospital. Working paper, ISDS Department, Louisiana State University.

Kim, D. 2005. An integrated supply chain management system: A case study in healthcare sector. *Lect Notes Comput Sci* 3590:218–227.

Krishnan, K., and V. Rao. 1965. Inventory control in N warehouses. *J Ind Eng* 16(3): 212–215.

Landry, S., and M. Beaulieu. 2000. *Étude internationale des meilleures pratiques de logistique hospitalière*. Montreal: Cahier de recherche du groupe CHAINE.

Lapierre, S., and A. Ruiz. 2007. Scheduling logistic activities to improve hospital supply systems. *Comput Oper Res* 34(3):624–641.

Lee, H., and J. Wu. 2006. A study on inventory replenishment policies in a two-echelon supply chain system. *Comput Ind Eng* 51(2):257–263.

Li, L., and W. Benton. 1996. Performance measurement criteria in health care organizations: Review and future research directions. *Eur J Oper Res* 93(3):449–468.

Lian, Z., and L. Liu. 2001. Continuous review perishable inventory systems: models and heuristics. *IIE Trans* 33(9):809–822.

Liu, L., and Z. Lian. 1999. (s, S) continuous review models for products with fixed lifetimes. *Oper Res* 47(1):150–158.

Lunn, T. 2000. Ways to reduce inventory. *Hosp Mater Manag Q* 21(4):1–7.

Marmor, T., and K. Okma 1998. Cautionary lessons from the West: What (not) to learn from other countries' experience in the financing and delivery of health care. In *The state of social welfare*, edited by P. Flora, P. R. de Jong, J. Le Grand, and J. Y. Kim, 327–350. Aldershot, UK: Ashgate.

McKinsey and Company. 2004. Will health benefit costs eclipse profits. *The McKinsey Quarterly Chart Focus Newsletter* (September 2004).

McKinsey Global Institute. 2007. Accounting for the cost of health care in the United States. http://www.mckinsey.com/mgi/rp/healthcare/accounting_cost_healthcare.asp (accessed October 15, 2008).

Meijboom, B., and B. Obel. 2007. Tactical coordination in a multi-location and multi-stage operations structure: A model and a pharmaceutical company case. *Omega* 35(3):258–273.

Michelon, P., M. Dib Cruz, and V. Gascon. 1994. Using the tabu search method for the distribution of supplies in a hospital. *Ann Oper Res* 50(1):427–435.

Minner, S., and E. Silver. 2005. Multi-product batch replenishment strategies under stochastic demand and a joint capacity constraint. *IIE Trans* 37(5):469–479.

Minner, S., and E. Silver. 2007. Replenishment policies for multiple products with compound-Poisson demand that share a common warehouse. *Int J Prod Econ* 108(1-2): 388–398.

Naddor, E. 1975. Optimal and heuristic decisions on single- and multi-item inventory systems *Mgmt Sci*. 21(11):1234–1249.

Nahmias, S. 1975 Optimal ordering policies for perishable inventory—II. *Oper Res* 23(4):735–749.

Nahmias, S. 1982. Perishable inventory theory: A review. *Oper Res* 30(4):680–707.

Nahmias, S., and S. Smith. 1994. Optimizing inventory levels in a two-echelon retailer system with partial lost sales. *Manag Sci* 40(5):582–596.

Nicholson, L., A. J. Vakharia, and S. Selcuk Erenguc. 2004. Outsourcing inventory management decisions in healthcare: Models and application. *Eur J Oper Res* 154(1):271–290.

Ohno, K., and T. Ishigaki. 2001. A multi-item continuous review inventory system with compound Poisson demands. *Math Meth Oper Res* 53(1):147–165.

Organization for Economic Co-operation and Development (OECD). 1994. *Health: Quality and choice*. Paris: OECD, OECD Health Policy Studies No. 4.

Organization for Economic Co-operation and Development (OECD). 1995. *New directions in health care policy*. Paris: OECD, OECD Health Policy Studies No. 7.

Organization for Economic Co-operation and Development (OECD). 1996. *Health care reform: The will to change*. Paris: OECD, OECD Health Policy Studies No. 8.

Perini, V., and L. Vermeulen Jr. 1994. Comparison of automated medication-management systems. *Am J Hosp Pharm* 51(15):1883–1891.

Pitta, D., and M. Laric. 2004. Value chains in health care. *J Consum Market* 21(7):451–464.

Plunkett Research Group. 2008. Health care trends. http://www.plunkettresearch.com/Industries/HealthCare/HealthCareTrends/tabid/294/Default.aspx (accessed September 10, 2008).

Poisal, J., C. Truffer, S. Smith, A. Sisko, C. Cowan, S. Keehan, B. Dickensheets, and the National Health Expenditure Accounts Projections Team. 2007. Health spending projections through 2016: Modest changes obscure part D's impact. *Health Aff* 21(2): W242–253.

Porter, M. 1985. *Competitive advantage: Creating and sustaining superior performance*. New York: The Free Press.

Prashant, N. 1991. A systematic approach to optimization of inventory management functions. *Hosp Mater Manag Q* 12(4):34–38.

Prosser, H., and T. Walley. 2005. A qualitative study of GPs' and PCO stakeholders' views on the importance and influence of cost on prescribing. *Soc Sci Med* 60(6):1335–1346.

Prosser, H., and T. Walley. 2007. Perceptions of the impact of primary care organizations on GP prescribing: The iron fist in the velvet glove? *J Health Organ Manage* 21(1):5–26.

Rivard-Royer, H., S. Landry, and M. Beaulieu. 2002. Hybrid stockless: A case study. *Int J Oper Prod Manag* 22(4):412–424.

Roberts, D. 1962. Approximations to optimal policies in a dynamic inventory model studies. In *Applied probability and management science*, edited by K. Arrow, S. Kardin, and H. Scarf, 207–229. Stanford, CA: Stanford University Press.

Rogers, D., and S. Tsubakitani. 1991. Inventory positioning/partitioning for backorders optimization for a class of multi-echelon inventory problems. *Decis Sci* 22(3):536–558.

Rothgang, H., M. Cacace, S. Grimmeisen, and C. Wendt. 2005. The changing role of the state in healthcare systems. *Eur Rev* 13(S1):187–212.

Sani, B., and B. Kingsman. 1997. Selecting the best periodic inventory control and demand forecasting methods for low demand items. *J Oper Res Soc* 48(7):700–713.

Schmidt, C., and S. Nahmias. 1985. (S–1, S) policies for perishable inventory. *Manag Sci* 31(6):719–728.

Seo, Y., S. Jung, and J. Hahm. 2001. Optimal reorder decision utilizing centralized stock information in a two-echelon distribution system. *Comput Oper Res* 29(2):171–193.

Simpson Jr., K. 1959. A theory of allocations of stocks to warehouses. *Oper Res* 7(6): 797–805.

Sinha, D., and K. Matta. 1991. Multiechelon (R, S) inventory model. *Decis Sci* 22(3): 484–499.

Smith-Daniels, V. 2006. Supply chains critical to well-being of health-care systems. *Knowledge@W.P. Carey*, http://knowledge.wpcarey.asu.edu/article.cfm?articleid=1245 (accessed October 14, 2008).

Syntetos, A., and J. Boylan. 2006. On the stock control performance of intermittent demand estimators. *Int J Prod Econ* 103(1):36–47.

Tarabusi, C., and G. Vickery. 1998a. Globalization in the pharmaceutical industry, part I. *Int J Health Serv* 28(1):67–105.

Tarabusi, C., and G. Vickery. 1998b. Globalization in the pharmaceutical industry, part II. *Int J Health Serv* 28(2):281–303.

van Merode, G., S. Groothuis, and A. Hasman. 2004. Enterprise resource planning for hospitals. *Int J Med Inf* 73(6):493–501.

Veinott, A. 1965. Optimal policy for a multiproduct, dynamic nonstationary inventory problem. *Manag Sci* 12:206–222.

Veral, E., and H. Rosen. 2001. Can a focus on costs increase costs? *Hosp Mater Manag Q*
22(3):28–35.

Vissers, J., J. Bertrand, and G. de Vries. 2001. A framework for production control in health
care organizations. *Prod Plann Contr* 12(6):591–604.

Weiss, H. 1980. Optimal ordering policies for continuous review perishable inventory models.
Oper Res 28(2):365–374.

Zhu, L., H. Wang, and L. Zhao. 2005. Comparative research on three-echelon and two-echelon
medicine inventory model with positive lead-time. *J Southeast Univ* (English Edition)
21:500–505.

6 Inventory Modeling for Complex Emergencies in Humanitarian Relief Operations

Benita M. Beamon and Stephen A. Kotleba
University of Washington
Seattle, Washington

CONTENTS

6.1   INTRODUCTION

In 1998 there were 400 natural disasters reported in the International Federation of the Red Cross World Disasters Report (1999). These natural disasters affected more than 144 million people, resulting in 90,000 deaths and 5 million temporarily displaced persons. In 1998, the governmental donor community spent more than \$3 billion responding to the immediate effects of natural disasters and man-made emergencies (International Federation of Red Cross and Red Crescent Societies 1999). In 2005, the International Federation of the Red Cross World Disasters Report stated that the number of disasters had risen dramatically to an average of 707 disasters per year from 1999 to 2003, affecting an average of 213 million people per year (International Federation of Red Cross and Red Crescent Societies 2005).

The challenge for nongovernmental organizations (NGOs) and other relief agencies responding to global emergencies (e.g., volcanic eruptions, earthquakes, floods, war) is how to prepare and manage relief activities in an unpredictable environment. Even with modern scientific advances, the onset of large-scale disasters can occur with little warning, and once a disaster strikes, the relief response often necessitates a variety of NGO expertise. Regardless of an NGO's operational specialty (e.g., housing and shelter, water and sanitation, pharmaceuticals), and whether the response is on a national or international level, each NGO shares the common critical objective of rapidly delivering the correct amount of goods, people, and monetary resources to the needed locations. As a result, the ability of an NGO's supply chain and logistic operations directly affects the success of a relief effort.

Recent studies such as Fenton (2003) and Thomas (2003) often compare the current state of supply chain management capabilities within humanitarian organizations to that of the commercial sector in the 1970s and '80s. At that time, the commercial sector began to realize the strategic advantages and significant improvements supply chain management could offer in effectiveness and efficiency. This led to extensive research in the area of supply chain and logistical analysis. Even so, quantitative methods and principles are rarely applied to humanitarian operations. Also, as global relief responses are growing in scale and magnitude, humanitarian supply chain management is becoming increasingly complex. The growing logistical needs are outpacing the capabilities of current management approaches. This is partly due to the aid sector's regard for logistics as a necessary expense (rather than an important strategic component in their work), the lack of depth in operational knowledge (due to high humanitarian agency employee turnover), and the general lack of investment in technology and communication. The trends of increasing numbers of natural disasters and increasing emphasis on accountability have served as motivating factors to increase understanding and improve the capacity for delivering humanitarian relief. This work focuses specifically on developing an inventory management strategy for a warehouse supporting a complex emergency relief operation.

6.2   LITERATURE REVIEW

Quantitative tools for emergency relief have typically come in the form of mathematical and network flow models. Oh and Haghani (1996) analyze the transportation of large amounts of many different commodities, such as food, clothing, medicine, medical supplies, machinery, and personnel in the most efficient manner to minimize the loss of life and maximize the efficiency of the rescue operations. The result is a formulation and solution of a multicommodity, multimodal network flow model for a generic disaster relief operation. The model is constructed to have four nodes, five arcs, and three transportation modes. Two heuristics are developed with the first utilizing a Lagrangian relaxation approach and the second employing an iterative fix-and-run process. Oh and Haghani (1997) further explore their heuristic models and provide more detailed analysis.

As has been well documented in disaster relief analysis, the infrastructure for the transportation of relief supplies is often unreliable. Due to this unreliability, Barbarosoglu et al. (2002) abandon dependence on road networks and focus on the use of helicopters for aid delivery and rescue missions during natural disasters. Their research is mainly concerned with developing mathematical models that solve tactical and operational scheduling decisions regarding helicopter activities. The authors utilize existing research in helicopter routing to address crew assignment, routing, and transportation issues encountered during the initial response phase of disaster management. Infrastructure unreliability as an obstacle to developing and maintaining supply chains was also the motivation behind Thomas (2002). The author develops a method for quantifying the reliability of supply chains for contingency logistics systems based on reliability interference theory. Barbarosoglu and Arda (2004) further explore modeling the uncertainty involved in emergency response. They develop a two-stage stochastic programming framework for transportation planning in disaster response. In this work, the authors expanded on Oh and Haghani's (1996) deterministic multicommodity, the multimodal network flow problem to include uncertainties that exist in estimating resource requirements of first-aid commodities, the vulnerability of resource provider facilities, and the survivability of the connecting routes in the disaster area.

Ozdamar et al. (2004) examines logistics planning in emergencies involving dispatching commodities to distribution centers of affected areas. The network flow model developed by the author addresses a dynamic time-dependent transportation problem, and repetitively derives a solution at given time intervals to represent ongoing aid delivery. The model regenerates plans incorporating new requests for aid materials, new supplies, and transportation modes that become available during the current planning time horizon. The plan indicates the optimal mixed pick up and delivery schedules for vehicles within the considered planning time horizon as well as the optimal quantities and types of loads to be picked up and delivered on these routes.

Sakakibara et al. (2004) also examine commodity flow in a road network during disaster response, and partitions the road network into isolated components.

The authors use a topological index to quantify road network accessibility, and then evaluate the isolation of districts in a disaster location. The authors present a methodology for specifying effective road links for avoiding functional isolation of districts.

Mathematical modeling of inventory management in emergency relief efforts has received little attention in the literature. Depending on the nature of the disaster, humanitarian emergency operations can continue for many years. To ensure continuous capacity in long-term responses, NGOs often institute arrangements for the storage of relief items in warehouses located near the response location. This research focuses on developing an inventory management policy to improve the effectiveness and efficiency of emergency relief during long-term humanitarian responses.

6.3   SYSTEM TO BE MODELED: SOUTH SUDAN RELIEF EFFORTS

The last decade has seen a decisive increase in the loss of life, property, and material damage due to the rising occurrence of natural disasters (United States Agency for International Development 2005). Also on the rise and equally as devastating, are man-made disasters, which are referred to within humanitarian agencies as complex humanitarian emergencies (United States Committee for Refugees and Immigrants 2004). UNICEF (2003) defines a complex humanitarian emergency as, "A humanitarian crisis in a country, region, or society where there is significant or total breakdown of authority resulting from internal or external conflict and which requires an international response that extends beyond the mandate or capacity of any single humanitarian agency."

Complex humanitarian emergencies are typically rooted in racially, ethnically, or religiously charged warfare, and are frequently characterized by horrific violations of human rights. The disturbing trend with complex humanitarian emergencies is that when conflicts erupt within the borders of a country, the dividing line between civilians and combatants is frequently blurred. Militant or rebellious groups are usually the same civilians living and socializing in and around the villages they attack. This type of warfare commonly resorts to the use of insidious tactics where humanitarian agencies are denied access to groups of people in need of assistance. Also, at the center of these shocking developments is the emergence of civilians, including women, children, and humanitarian workers, as the deliberate targets of warfare rather than its incidental victims. The vast devastation of complex humanitarian emergencies can also be attributed to the typically long length of the disaster, which imposes not only financial hurdles, but also legal, moral, and political dilemmas to those attempting to provide relief. Humanitarian efforts responding to complex emergencies often last for many years, and therefore require logistics systems that can support long-term relief activities.

The destruction resulting from long-term complex humanitarian emergencies has recently been witnessed in the Balkans tragedy, the genocide in Rwanda, and most dramatically in the civil war in south Sudan. The United States Holocaust Memorial Museum (2005) describes the warfare in Sudan, Africa's largest country, as one of the most devastating humanitarian crises ever to affect the world. Since 1983, civil war has ravaged south Sudan, leaving over 2 million dead and over

4 million displaced (United States Committee for Refugees and Immigrants 2004). For over twenty years, humanitarian agencies have been responding to the emergency by airlifting aid to many parts of the underdeveloped south Sudan region. The United States Committee for Refugees and Immigrants World Refugee Survey: Sudan (2004) estimates that millions of people are dependent on the relief supplies provided by humanitarian agencies. Air access is vital for agencies providing relief as the existing roads in Sudan are often in poor condition, heavily mined, prone to attack by bandits and militia, and at times impassable due to seasonal rains (United States Holocaust Memorial Museum 2005). The south Sudan civil war has been fought primarily with guerrilla warfare tactics, which are based on principles of ambush and sabotage. The result is a sporadic need for humanitarian aid in unpredictable quantities. Prepositioning of relief supplies near the affected area has proven to be an effective strategy for responding to emergencies of this nature (UNICEF 2005).

In 1989, in response to the continual need for humanitarian aid in south Sudan, Operation Lifeline Sudan (OLS) was created under the coordination of the United Nations as a consortium of UNICEF, the World Food Programme, and more than thirty-five NGOs. The purpose of OLS was for the United Nations to provide the necessary air transport and security for NGO operations, and to provide a strict code of conduct to maintain high standards and impartiality in the delivery of humanitarian assistance to Sudanese civilians. OLS operates out of a United Nations base in the northwest Kenyan town of Lockichoggio, and provides more than a dozen daily airlifts of food, relief supplies, and people (Humanitarian and War Project 2005). The air transport and security provided by the OLS have allowed many NGOs to preposition relief items in warehouses throughout Lockichoggio. As such, Lockichoggio has developed into a logistical hub for the south Sudan relief efforts.

Global prepositioning of relief supplies is an expansion of the warehousing strategy typically seen in complex humanitarian emergency responses. This response strategy allows NGOs to respond quickly to disasters with relief supplies from strategically stocked warehouses throughout the world. Global prepositioning is a relatively new approach, and currently only a few NGOs can support the large expense of operating a warehouse that serves the international community. World Vision International has taken the lead in implementing a global prepositioning system. Their global prepositioning units (GPUs) are part of a strategy that allows them to respond rapidly and effectively to large-scale emergencies. The GPU system was initiated in 2000 with three warehouse locations (Denver, Colorado; Brindisi, Italy; Hanover, Germany), but the full impact of their operations has yet to be determined. The trade-off of operating a global prepositioned system is rapid response but large transportation costs. The challenge for NGOs is to integrate a GPU system into a long-term humanitarian relief response effectively and efficiently.

The objective here is to develop an efficient, quick-response warehouse inventory policy for a humanitarian organization responding to a complex humanitarian emergency. The analysis is based on a case study of a single humanitarian agency operating a warehouse in Lockichoggio and responding to the south Sudan crisis within the OLS framework.

6.4 HUMANITARIAN RELIEF INVENTORY MODEL

Quantitative multisupplier supply chain inventory modeling is an active area of research (see Minner and von-Guericke [2003] for a review of such models). Our research develops a multisupplier inventory model that accounts for the unique demand patterns that occur in humanitarian emergency relief operations. We will first provide the relevant model framework and then describe our model in detail. Our model develops an inventory policy for the unique characteristics of a long-term complex humanitarian emergency relief response, considering the specific characteristics of the ordering process and demand distribution, yielding new expressions for on-hand inventory, stockout probability, expected number of back orders, and ultimately the decision variables Q_1, r_1, Q_2, and r_2.

General multisupplier inventory models, such as Moinzadeh and Nahmias (1988), assume a continuous review inventory system with two options for resupply. These models are an extension of the standard (Q, r) inventory policy that allows for two lot sizes (Q_1 and Q_2), and two reorder levels (r_1 and r_2). An order of size Q_1 is placed when the inventory position reaches r_1 (this is a regular, or normal reorder option). An emergency reorder option is an expedited order of size Q_2 placed when the inventory reaches a position of r_2 (where $r_1 > r_2$). The lead times for normal and emergency reorders are assumed constant in the model and are represented as τ_1 and τ_2, respectively (where $\tau_1 > \tau_2$). In addition to requiring a shorter lead time, the items ordered through the emergency resupply channel are also assumed to incur higher fixed and per-unit ordering costs than the normal orders. The rest of the notation is given next with units in parentheses, and will be used throughout the rest of the chapter.

6.4.1 SUMMARY OF MODELING NOTATION

- K_1: Fixed cost for placing a normal order ($)
- K_2: Fixed cost for placing an emergency order ($)
- c_1: Per-unit cost for a normal order ($)
- c_2: Per-unit cost for an emergency order ($)
- h: Inventory holding cost per item per unit time ($)
- π: Back-order cost per item per unit time ($)
- p: Probability of a stockout occurring within a cycle
- OH: On-hand inventory per cycle (number of items)
- BO: Number of back-ordered items per cycle (number of units)
- T: Expected cycle length (time units)
- μ: Expected demand rate (quantity per unit time)
- D: Annual demand (number of units)
- Q_1: Quantity of relief items in a normal reorder (units)
- Q_2: Quantity of relief items in an emergency reorder (units)
- r_1: Reorder level for normal orders (units)
- r_2: Reorder level for emergency orders (units)

In this section, we present a multisupplier inventory model for the south Sudan relief operations. Data was collected on site in Lockichoggio, Kenya, from warehouse manager interviews and warehouse stock-keeping cards during September 2004. We developed the Sudan humanitarian inventory model as a continuous inventory review system with two options for resupply. The model allows for two lot sizes (Q_1 and Q_2) and two reorder levels (r_1 and r_2). An order of size Q_1 is placed when the inventory position reaches r_1 (a regular, or normal resupply option). The expedited emergency resupply option is of size Q_2 and placed when the inventory reaches a position of r_2. The normal and emergency resupply options represent placing orders with a supplier in Nairobi and an international supplier, respectively. The international supplier may be directly affiliated with the relief agency, as with the GPUs of World Vision, or may have no affiliation at all.

Our process for developing the model begins by analyzing the expected reorder level for each cycle. Next, we focus on the on-hand inventory and expected number of back orders, compute the probability of a stockout per cycle and the total cost equations, and finally derive optimality conditions.

6.4.2 ASSUMPTIONS

These assumptions will apply to the humanitarian relief model:

- $r_1 > r_2$, in a given cycle, an emergency order will never be placed before a normal order.
- $\tau_1 > \tau_2$, an emergency order has a shorter lead time.
- The international supplier will not experience stockouts.
- Replenishment lead time for normal orders is eight days, and for simplification is assumed constant.
- Items ordered through the emergency resupply channel incur a higher fixed and per-unit ordering cost than normal orders.
- Replenishment lead time for emergency orders is two days, and due to relative low variation is assumed constant.
- $Q_1 > r_1$, an order quantity will be large enough to not automatically trigger an additional reorder.
- Demand $\sim U[1, b]$ units.
- Demand occurs in discrete ten-day intervals.

6.4.3 EXPECTED REORDER LEVEL PER CYCLE

Humanitarian emergency relief systems require rapid response. Therefore, we assume that the inventory position is under continuous review. However, since the requests for items occur in discrete time intervals, the inventory policy shares some characteristics with periodic review models.

For the inventory model, we define the random variables:

- R_e: Actual inventory level at the time when a reorder is placed, $R_e \leq r_1$
- I: Inventory position just before a request is made, $I > r_1$
- D: Demand $\sim U[1, b]$, where D and b are discrete, and $b \geq 1$
- I': Inventory position just after a request has been fulfilled, $I' \in \Re$

In the humanitarian relief model, undershoots are possible (an undershoot is the difference between the set inventory reorder level and the actual inventory level when a reorder is placed).

Let the current inventory level be at a position i just before a request for d items is placed. Then, after the request has been fulfilled, the inventory level will drop to a position of i'. Therefore, define i' as:

$$i' = i - d \tag{6.1}$$

If $i' \le r_1$, then a reorder must be placed, and the resulting reorder level, r_e, is $r_e = i'$. Let Y be the number of units below r_1 that the inventory level reaches when a reorder is placed in any given cycle. Then define the random variable Y as:

$$Y = \begin{cases} r_1 - I' & I' < r_1 \\ 0 & \text{otherwise} \end{cases} \tag{6.2}$$

and it follows that if $I' < r_1$ then:

$$R_e = r_1 - Y \tag{6.3}$$

Then the expected value of the current inventory when the order is placed, $E[R_e]$, follows from Equation 6.3 as:

$$E[R_e] = r_1 - E[Y] \tag{6.4}$$

We begin by analyzing the current inventory position (I) with a reorder level (r_1), just before a demand (D) is placed. There are two cases for I.

Case 1: If $i > r_1 + b$, then i' cannot drop to r_1 or lower after the next request for items since the demand, D, is uniformly distributed and $\max(d) = b$. In this case, $i' > r_1$.

Case 2: If $i \le r_1 + b$ then i' can drop to r_1 or lower after the next request for items. In this case, $i' \le r$.

Therefore, in determining the expected reorder level, $E[Re]$, we only have to consider cases where reorders are possible. From earlier, reorders are possible only when the current inventory level, i, is in the range $[r_1 + 1, r_1 + b]$. A plot of all possible values of Y where $I \in [r_1 + 1, r_1 + b]$ is given in Figure 6.1. The i values, which represent the current inventory position before an order is received, are plotted along the x-axis. y-Values, which are the number of units below the reorder level, r_1, when a reorder is placed, are plotted along the y-axis.

Now, let $h(y)$ be the probability density function of Y. Then, $h(y)$ is given by:

$$h(y) = \frac{b - y}{\sum_{n=1}^{b} n}, \quad 0 \le y \le b - 1$$

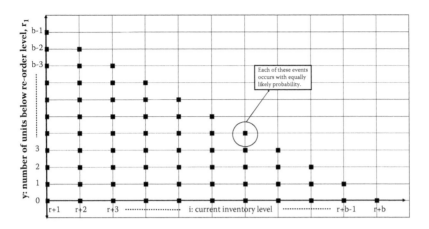

FIGURE 6.1 Distribution of Y.

Using the identity $\sum_{n=1}^{x} n = \frac{x^2+x}{2}$ yields:

$$h(y) = \frac{2(b-y)}{(b^2+b)}, \quad 0 \le y \le b-1 \tag{6.5}$$

We can then find the expected value of Y, E[Y], from the expression: $E[Y] = \sum_{y=0}^{b-1} y(h(y))$. It then follows that:

$$E[Y] = \sum_{y=0}^{b-1} \frac{2y(b-y)}{b^2+b} = \frac{2}{b^2+b}\left(\sum_{y=0}^{b-1} yb - \sum_{y=0}^{b-1} y^2\right)$$

Using the identity $\sum_{n=1}^{x} n^2 = \frac{(x^2+x)(2x+1)}{2}$, then

$$E[Y] = \frac{2}{b^2+b}\left(\frac{b((b-1)^2+(b-1))}{2} - \frac{((b-1)^2+(b-1))(2(b-1)+1)}{6}\right) = \frac{b}{3} - \frac{1}{3} \tag{6.6}$$

Substituting Equation 6.6 into Equation 6.4 gives the expected reorder level, E[R_e], as:

$$E[R_e] = r_1 - \left(\frac{b}{3} - \frac{1}{3}\right) \tag{6.7}$$

6.4.4 Expected Average Number of Units Held per Cycle

The fact that demands occur at discrete intervals has a significant effect on the amount of inventory held per cycle. This demand pattern causes the inventory level

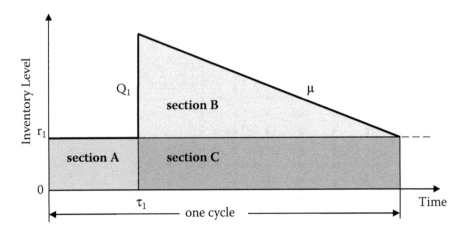

FIGURE 6.2 On-hand inventory plot for the humanitarian relief model with no emergency orders.

to experience periods of stability followed by sudden drops as items are removed in discrete batches. There are two distinct phases of the inventory cycle: Phase I and Phase II. Phase I is the replenishment lead time of an order of size Q_1. Phase II is the time between the receipt of an order of size Q_1 and the placement of the next order. As long as the lead time, τ_1, is less than the time between orders, then there will be no demand during the replenishment lead time, τ_1, and the inventory level will remain constant until the arrival of Q_1.

For simplification, we will approximate the discrete demand pattern of batches of items removed from inventory as continuous. The rate of demand is the expected demand per time, μ. The humanitarian relief inventory model is depicted in Figure 6.2.

6.4.5 Expected Average Number of Units Held per Cycle (No Emergency Orders)

The first situation to consider when determining the average amount of on-hand inventory is in cycles with no emergency orders. When no emergency orders occur, the inventory level during Phase I of the humanitarian relief model remains constant.

By dividing the graph in Figure 6.2 into three sections (section A, section B, section C), we can compute the total average on-hand inventory as the sum of each section. The area of section A is $r_1 \times \tau_1$, the area of section B is $((1/2) \times (Q_1/\mu) \times Q_1)$, and the area of section C is $((Q_1/\mu) \times r_1)$. The expression for OH_1 is the sum of these three terms and is given by:

$$OH_1 = \frac{Q_1^2}{2\mu} + \frac{Q_1 r_1}{\mu} + r_1 \tau_1 \tag{6.8}$$

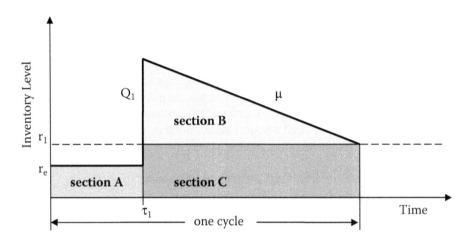

FIGURE 6.3 On-hand inventory depicting the expected reorder level, r_e.

Recall that when a reorder is placed, it is likely that the resultant inventory level will not be exactly equal to the reorder level, r_1, but at the actual reorder level, r_e, where $r_1 > r_e$ (see Figure 6.3).

We can again divide the total area under the curve of Figure 6.4 into three sections (section A, section B, section C) and compute the total on-hand inventory as the sum of each section. The area of section A is now $r_e \times \tau_1$, the area of section B is $((1/2) \times (r_e + Q_1 - r_1/\mu) \times (r_e + Q_1 - r_1))$, and the area of section C is $((r_e + Q_1 - r_1/\mu) \times r_1)$. OH_1 is the sum of these three terms and is given by:

$$OH_1 = \frac{(R_e + Q_1 - r_1)^2}{2\mu} + \frac{(R_e + Q_1 - r_1)r_1}{\mu} + \tau_1 R_e \qquad (6.9)$$

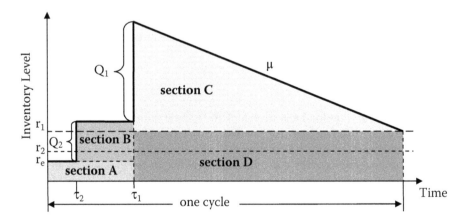

FIGURE 6.4 $OH_2{}^*$, a cycle in which an emergency order is placed with existing inventory.

6.4.6 EXPECTED AVERAGE NUMBER OF UNITS HELD PER CYCLE WITH EMERGENCY ORDERS

The second situation to consider when determining the average amount of on-hand inventory is cycles with emergency orders. Before we derive an expression for the average on-hand inventory when emergency orders are placed, we must first discuss the nature of a cycle with emergency orders. In the humanitarian relief model, an emergency order may be placed any time after a normal order is placed. There are two possible cases for the inventory level when an emergency order is placed:

Case 1: The inventory level is greater than or equal to zero ($i' \geq 0$).
Case 2: The inventory level has fallen below zero ($i' < 0$).

Figure 6.4 depicts Case 1, a cycle in which an emergency order is placed when the inventory level is greater than zero. We will refer to the on-hand inventory in this case as OH_2^*.

To determine OH_2^* we must find the area under the curve in Figure 6.4. By dividing the curve into sections, we can take the same approach as before. The area of section A is $r_e \times \tau_1$, the area of section B is $Q_1*(\tau_1 - \tau_2)$, the area of section C is $((1/2) \times ((r_e + Q_1 + Q_2 - r_1)/\mu) \times (r_e + Q_1 + Q_2 - r_1)$, and the area of section D is $(((r_e + Q_1 + Q_2 - r_1)/\mu) \times r_1)$. OH_2^* is the sum of these four terms and is given by:

$$OH_2^* = \frac{(R_e + Q_1 + Q_2 - r_1)^2}{2\mu} + \frac{(R_e + Q_1 + Q_2 - r_1)r_1}{\mu} + Q_1*(\tau_1 - \tau_2) + \tau_1 R_e \quad (6.10)$$

6.4.7 CYCLE LENGTH FOR RELIEF CYCLES WITH AND WITHOUT EMERGENCY ORDERS

Now we can determine expressions for the cycle length T_1 (cycles without emergency orders) and for the cycle length T_2 (cycles with emergency orders). Equation 6.8 is the on-hand inventory for a cycle with no emergency orders. The total cycle length is the sum of the normal order lead time (Phase I) plus the expected length of time between a normal order arrival and the placement of the next normal order (Phase II). The expected cycle length for cycles with no emergency orders, T_1, is given as:

$$T_1 = \tau_1 + \frac{(R_e + Q_1 - r_1)}{\mu} \quad (6.11)$$

Equation 6.10 is the on-hand inventory for a cycle with emergency orders. As before, the total cycle length is the sum of the normal order lead time (Phase I) plus the expected length of time between a normal order arrival and the placement of the next normal order (Phase II). The expected cycle length for cycles with an emergency order, T_2, is then given as:

$$T_2 = \tau_1 + \frac{(R_e + Q_1 + Q_2 - r_1)}{\mu} \quad (6.12)$$

6.4.8 REVISITING: AVERAGE NUMBER OF UNITS HELD PER CYCLE WITH EMERGENCY ORDERS

We can now compute the average on-hand inventory per unit time for the cases of placing and not placing an emergency order to test whether it is ever cost-effective to place an emergency order while inventory still exits. Dividing Equation 6.9 and Equation 6.10 by their respective average cycle lengths, Equation 6.11 and Equation 6.12, we obtain expressions for the average on-hand inventory per unit time without emergency orders, AOH_1, and with emergency orders, AOH_2^*:

$$AOH_1 = \frac{\frac{(R_e+Q_1-r_1)^2}{2\mu} + \frac{(R_e+Q_1-r_1)r_1}{\mu} + \tau_1 R_e}{\frac{R_e+Q_1-r_1}{\mu} + \tau_1} \tag{6.13}$$

$$AOH_2^* = \frac{\frac{(R_e+Q_1+Q_2-r_1)^2}{2\mu} + \frac{(R_e+Q_1+Q_2-r_1)r_1}{\mu} + Q_1 * (\tau_1 - \tau_2) + \tau_1 R_e}{\frac{R_e+Q_1+Q_2-r_1}{\mu} + \tau_1} \tag{6.14}$$

It can be shown that Equation 6.13 will always be less than Equation 6.14, and therefore it is never economical to place an emergency order when the current inventory level is greater than zero.

This does not mean that placing an emergency order will never be beneficial. We have just shown that at the time of a reorder, if the inventory level is greater than or equal to zero, than it is not economical to place an emergency order, as it will only increase holding costs. However, if the inventory has fallen below zero (a stockout occurs), then supplies have been back-ordered and an outstanding demand remains in the field. In this situation, it is reasonable to place an emergency order. Instead of waiting for the replenishment lead time of τ_1, an emergency order can be placed and supplies delivered to the field within the time of τ_2. Also, as we have shown that any time the emergency order creates a positive inventory level, our holding costs increase. Therefore, when an emergency order is placed, it should be for exactly the amount of supplies back-ordered, which would bring our current inventory level to zero. Hence, for the humanitarian relief model we set Q_2 = expected number of back orders (E[BO]). This cycle is depicted in Figure 6.5.

The area under the curve in Figure 6.5 represents the amount of on-hand inventory during a cycle with an emergency order. Since the amount back-ordered is never considered part of our on-hand inventory, our reorder level, R_e, is reduced to zero. Substituting $R_e = 0$ into Equation 6.10 gives the following expression for on-hand inventory during a cycle with an emergency order:

$$OH_2 = \frac{(Q_1 - r_1)^2}{2\mu} + \frac{(Q_1 - r_1)r_1}{\mu} \tag{6.15}$$

6.4.9 TOTAL AVERAGE NUMBER OF UNITS HELD PER CYCLE

The total average on-hand inventory per cycle, OH, considers the probability of cycles with and without emergency orders. We define p as the probability that a cycle

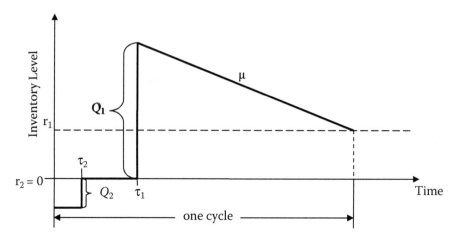

FIGURE 6.5 Realization of a cycle in the humanitarian relief model with an emergency order.

will have an emergency order. Then OH is given by:

$$OH = (1-p)OH_1 + pOH_2 \qquad (6.16)$$

Substituting the average on-hand inventory for cycles without emergency orders, Equation 6.9, and the average on-hand inventory for cycles with emergency orders, Equation 6.15, into Equation 6.16 yields the following approximation for the average on-hand inventory during a cycle:

$$OH = (1-p)OH_1 + pOH_2 = (1-p)\left(\frac{(r_e + Q_1 - r_1)^2}{2\mu} + \frac{(r_e + Q_1 - r_1)r_1}{\mu} + \tau_1 r_e \right)$$

$$+ p\left(\frac{(Q_1 - r_1)^2}{2\mu} + \frac{(Q_1 - r_1)r_1}{\mu} \right)$$

$$= r_e(1-p)\left(\frac{Q_1}{\mu} + \tau_1 \right) + \left(\frac{r_e^2(1-p) + Q_1^2 - r_1^2}{2\mu} \right) \qquad (6.17)$$

6.4.10 Total Average Cycle Length

As in the calculation of the total average on-hand inventory in Equation 6.16, the calculation for the total average cycle length considers the probability of both cycles with and without emergency orders. The total cycle length, T, is:

$$T = (1-p)T_1 + pT_2 \qquad (6.18)$$

The expected cycle length for cycles with an emergency order, T_2, as shown in Figure 6.5, is:

$$T_2 = \tau_1 + \frac{(Q_1 - r_1)}{\mu} \qquad (6.19)$$

To determine the total expected cycle length, we substitute Equation 6.15 and Equation 6.11 into Equation 6.14, which simplifies to:

$$T = \tau_1 + \frac{r_e(1 - p) + Q_1 - r_1}{\mu} \qquad (6.20)$$

6.4.11 EXPECTED NUMBER OF BACK ORDERS PER CYCLE

To determine the total cost per cycle due to back orders, we must first determine the expected number of back orders per cycle, E[BO]. Calculating E[BO] is very similar to calculating the expected reorder level, E[R_e]. In determining E[BO], we now consider the number of units below zero the inventory level reaches at the time a reorder is placed (that is, when y, the number of units below the reorder level, r_1, is greater than r_1, ($y > r_1$)). Define Y' as the number of units below zero the inventory reaches when a reorder is placed. Then, $Y' = Y - r_1$, and the number of back orders per cycle, BO, is given by $BO = Y'$, and E[BO] = E[Y'].

In calculating E[R_e], we determined the probability density function of Y, and that the range of Y was defined on y [0, b – 1], where $y = 0$ describes a situation where the inventory position falls to exactly r_1 when a reorder is placed. Therefore, since we are now interested in the situation in which $y > r_1$, our range for Y' is defined as $Y' \in [r_1 + 1, b - 1]$ as shown in Figure 6.6.

Since Y' is a subset of Y, we can take the probability density function of Y, originally given in Equation 6.5, and limit its range to $r_1 + 1 \le y \le b - 1$. Substituting $y = y' + r_1$ into Equation 6.5 and restricting the range allow us to compute the

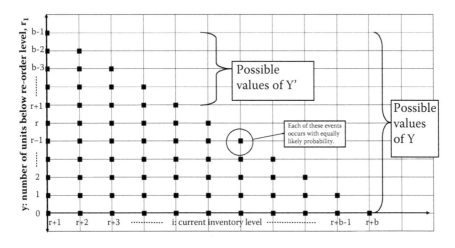

FIGURE 6.6 Probability density function of Y and Y'.

probability density function of Y', $h(y')$, as:

$$h(y') = \frac{2(b-r-y')}{(b^2+b)}, \quad 1 \le y' \le b - r_1 - 1 \qquad (6.21)$$

Then, $E[Y'] = \sum_{y'=1}^{b-r_1-1} y'(h(y'))$, which can be computed as:

$$E[Y'] = \sum_{y'=1}^{b-r_1-1} y'(h(y'))$$

$$= \frac{2}{b^2+b}\left(\sum_{y=1}^{b-r_1-1} y'b - \sum_{y=1}^{b-r_1-1} y'r - \sum_{y=1}^{b-r_1-1} y^2 \right)$$

$$= \frac{2}{b^2+b}\left(\frac{b(\eta^2+\eta)}{2} - \frac{r_1(\eta^2+\eta)}{2} - \frac{(\eta^2+\eta)(2\eta+1)}{6} \right)$$

$$= \frac{b^3 - 3b^2r_1 + 3br_1^2 - r_1^3 - b + r_1}{3(b^2+b)}$$

where $\eta = b - r_1 - 1$. Therefore, the expected number of back orders per cycle, $E[BO]$, is:

$$E[BO] = \frac{b^3 - 3b^2r_1 + 3br_1^2 - r_1^3 - b + r_1}{3(b^2+b)} \qquad (6.22)$$

6.4.12 Per-Cycle Back Order Probabilities

Previously, we defined p as the probability of an emergency order occurring within a cycle (and therefore a stockout occurring within a cycle). We can now derive a value for p. We also defined Y' as the number of units below zero the inventory reaches when a reorder is placed, which is the number of back orders, and derived the probability density function of Y', $h(y')$, in Equation 6.5. Therefore, p, the probability of a stockout in a cycle, is the sum of the individual probability densities of y' where $y' \ge 1$. Therefore:

$$p = \sum_{y'=1}^{b-r_1-1} h(y')$$

$$= \sum_{y'=1}^{b-r_1-1} \frac{2(b-r_1-y')}{(b^2+b)}$$

$$= \frac{2}{(b^2+b)}\left[\sum_{y'=1}^{b-r_1-1} b - \sum_{y'=1}^{b-r_1-1} r_1 - \sum_{y'=1}^{b-r_1-1} y' \right]$$

$$= \frac{2}{(b^2 + b)}\left[b(b - r_1 - 1) - r_1(b - r - 1) - \left(\frac{(b - r_1 - 1)^2 + (b - r_1 - 1)}{2}\right)\right]$$

$$= \frac{b^2 - 2br_1 + r_1^2 - b + r_1}{(b^2 + b)} \qquad (6.23)$$

6.4.13 Total Cost Equation

The total cost is computed on a per-cycle basis, where one cycle is defined as the time between consecutive orders of Q_1. The total cost is the sum of the fixed cost of placing a normal order, K_1; the per-unit cost for a normal order, $c_1 Q_1$; the fixed and per-unit cost of an emergency order multiplied by the probability of placing an emergency order, $p(K_2 + c_2 Q_2)$; the inventory holding costs, $h(OH)$; and the back-order costs, $\pi E[BO]$.

Therefore, the total cost in a cycle, TC, is given as:

$$TC(Q_1, Q_2, r_1, r_2) = K_1 + c_1 Q_1 + p(K_2 + c_2 Q_2) + h[OH] + \pi E(BO) \qquad (6.24)$$

It then follows that the average cost per unit time, AC, is:

$$AC(Q_1, Q_2, r_1, r_2) = \frac{TC(Q_1, Q_2, r_1, r_2)}{T} \qquad (6.25)$$

6.4.14 Deriving Optimal Conditions

W can now derive optimal conditions. For fixed values of r_1 and r_2, the optimal values of Q_1 and Q_2 can be found using the optimality conditions for the average cycle cost. The optimality conditions are:

$$\frac{\partial AC}{\partial Q_1} = \frac{\partial AC}{\partial Q_2} = 0 \qquad (6.26)$$

Recall that an emergency order of Q_2 is placed to raise the inventory level back to zero. Therefore, Q_2 is the expected number of back orders in a cycle, so $Q_2 = E[BO]$. Substituting $Q_2 = E[BO]$ into the total cost equation, Equation 6.24, gives:

$$TC(Q_1, r_1, r_2) = K_1 + c_1 Q_1 + p(K_2 + c_2 E[BO]) + h[OH] + \pi E[BO] \qquad (6.27)$$

Since the expected number of back orders (Equation 6.22) and the total average on-hand inventory (Equation 6.17) are independent of Q_2, the total cost equation (Equation 6.24) is independent of Q_2. The average total cycle length (Equation 6.20) is also independent of Q_2, and therefore the partial derivative in Equation 6.26 with respect to Q_2 is equal to zero. This reduces the number of optimality equations to one, which can be solved directly for Q_1. The optimality condition is as follows:

$$\frac{dAC}{dQ_1} = 0$$

where

$$\frac{dAC}{dQ_1} = -\frac{TC}{T^2}\frac{dT}{dQ_1} + \frac{1}{T}\frac{dTC}{dQ_1} \tag{6.28}$$

Solving Equation 6.28 gives the following closed-form expression for Q_1 (see Appendix for derivation):

$$Q_1 = r_1 + r_e(1-p) - \tau_1\mu$$

$$\pm\frac{1}{h}\sqrt{\begin{array}{l}h^2(r_e(2\tau_1\mu(1-p) + pr_e(1-p)) + \tau_1\mu(2r_1 + \tau_1\mu)) \\ + 2h\mu(c_1(r_e(1-p) + r_1 - \mu\tau_1) + K_1 + pK_2 + E(BO)(pc_2 + \pi))\end{array}} \tag{6.29}$$

6.4.15 SOLUTION PROCEDURE

The following steps outline the procedure for developing a solution to the model:

1. Select an order stockout risk (OSOR). The OSOR is the risk we are willing to accept that any cycle in our system will experience a stockout, and will depend on the severity of the emergency. Intuitively, the more critical the aid, the lower the OSOR we will select. We can set the OSOR as we wish since the stockout probability, Equation 6.23, is independent of Q_1.
2. Set $p = $ OSOR and determine the reorder level r_1 from Equation 6.23.
3. Use r_1 to determine the expected reorder level, $E[R_e]$, from Equation 6.4.
4. Determine the expected number of back orders per cycle, $E[BO]$, using r_1 in Equation 6.22.
5. Substitute r_1, r_e, p, $E[BO]$ and the initial parameters into Equation 6.29 and solve for Q_1.

6.5 CONCLUSIONS AND FUTURE RESEARCH

In this research, our objective was to develop an inventory model for a prepositioned warehouse responding to a complex humanitarian emergency, which is one of the three emergency classifications to which humanitarian and nongovernmental agencies respond (rapid onset and slow onset are the other two). Complex humanitarian emergencies are unique due to their unpredictable demand patterns and long durations. The high (often life-threatening) stakes of humanitarian relief place a heavy emphasis on a quick logistics response. It is essential that humanitarian logistics operations (including warehousing in a complex emergency) be performed as efficiently and effectively as possible. Field research for this chapter was performed with our host organization, World Vision International. Data was collected from its warehouse operations in Lockichoggio, Kenya, as it was responding to the complex humanitarian emergency in south Sudan. Based on our recorded data, we developed a mathematical model that optimized the reorder quantity and reorder level based on reordering, holding, and back-order costs.

The research presented in this chapter is a first step in developing strategic inventory management systems for humanitarian relief. The model investigated a single "item" (which may be interpreted as a single type of relief kit or a single set of items), developed order quantities that were independent of vehicle or container sizes, and assumed a continuous demand approximation. Future work would investigate the effects of a (correlated) multi-item system that incorporates container sizes, which would likely prove especially important in a relief situation. Another next step would include explicit modeling of the discrete inventory removals (rather than using a continuous approximation). Additionally, a more thorough study to understand the full implications of back-order costs within humanitarian logistics is needed. The back-order cost represents a penalty for unmet demand, but since there are no financial profits made in humanitarian logistics, the lost sales approach used in commercial logistics is not necessarily appropriate. A back-order cost in humanitarian logistics represents potential suffering (or loss of life) endured by a potential recipient from not receiving a relief item, but also for the "advertising" opportunity lost by the relief agency for not delivering an item. NGOs do not have an income stream to fund their work outside of donations and therefore must remain vigilant that their actions are being noticed within the donor community. The back-order cost used in our model is a one-time penalty applied to each unit of demand that could not be immediately met from inventory. Eventually, back orders are met from an expedited order, but the length of time required to fulfill back orders is not considered. In the real system, there is a window of opportunity during which relief items can be used in the field, and the longer the delay for a needed item, the more devastating the effects. Future research would analyze the impact of time-dependent shortage costs and develop a more detailed quantification methodology for assessing shortages in humanitarian logistics.

ACKNOWLEDGMENTS

This work was partly supported by the Interdisciplinary Program in Humanitarian Relief at the University of Washington. The authors would also like to thank World Vision International, and especially Randy Strash and George Fenton for their support and effort during the data gathering process, and to the anonymous referees for their helpful comments and suggested improvements.

APPENDIX: CLOSED-FORM EXPRESSION FOR Q_1

Given the optimality condition

$$\frac{dAC}{dQ_1} = 0$$

where

$$AC(Q_1, Q_2, R_1, R_2) = TC(Q_1, Q_2, R_1, R_2)/T,$$

$$TC(Q_1, Q_2, r_1, r_2) = K_1 + c_1Q_1 + p(K_2 + c_2E[BO]) + h[OH] + \pi E[BO],$$

$$T = \tau_1 + \frac{R_e(1-p) + Q_1 - r_1}{\mu}$$

and

$$OH = r_e(1-p)\left(\frac{Q_1}{\mu} + \tau_1\right) + \left(\frac{r_e^2(1-p) + Q_1^2 - r_1^2}{2\mu}\right)$$

we can form the expression for AC as:

$$AC = \frac{K_1 + c_1 Q_1 + p(K_2 + c_2 E(BO)) + h\left(R_e(1-p)\left(\frac{Q_1}{\mu} + \tau_1\right) + \frac{R_e^2(1-p) + Q_1^2 - r_1^2}{2\mu}\right) + \pi E(BO)}{\tau_1 + \frac{R_e(1-p) + Q_1 - r_1}{2\mu}}$$

We can factor and simplify AC to get the following:

$$AC = \frac{-1}{2(\tau_1 \mu + R_e p + Q_1 - r_1)}$$

$$\left(-2K_1 \mu - 2c_1 Q_1 \mu - 2p\mu K_2 - 2p\mu c_2 E(BO) - 2hR_e Q_1 - 2hR_e \tau_1 \mu \right.$$

$$\left. + 2hR_e p Q_1 + 2hR_e p \tau_1 \mu - hR_e^2 + hR_e^2 p - hQ_1^2 + hr_1^2 - 2\pi E(BO)\mu\right)$$

Taking the derivative of AC with respect to Q_1 gives:

$$\frac{dAC}{dQ_1} = -\frac{-2c_1 \mu - 2hR_e + 2hR_e p - 2hQ_1}{2(\tau_1 \mu + R_e - R_e p + Q_1 - r)} + \frac{1}{2(\tau_1 \mu + R_e - R_e p + Q_1 - r_1)^2}$$

$$\left(-2K_1 \mu - 2c_1 Q_1 \mu - 2p\mu K_2 - 2p\mu c_2 E(BO) - 2hR_e Q_1 - 2hR_e \tau_1 \mu + 2hR_e p Q_1 \right.$$

$$\left. + 2hR_e p \tau_1 \mu - hR_e^2 + hR_e^2 p - hQ_1^2 + hr_1^2 - 2\pi E(BO)\mu\right)$$

Setting $\frac{dAC}{dQ_1} = 0$ yields the following expression:

$$\frac{-2c_1 \mu - 2hr_e + 2hr_e p - 2hQ_1}{2(\tau_1 \mu + r_e - r_e p + Q_1 - r)} = \frac{1}{2(\tau_1 \mu + r_e - r_e p + Q_1 - r)^2}$$

$$\left(-2K_1 \mu - 2c_1 Q_1 \mu - 2p\mu K_2 - 2p\mu c_2 E(BO)\right.$$

$$- 2hr_e Q_1 - 2hr_e \tau_1 \mu + 2hr_e p Q_1 + 2hr_e p \tau_1 \mu$$

$$\left. - hr_e^2 + hr_e^2 p - hQ_1^2 + hr^2 - 2\pi E(BO)\mu\right)$$

Solving for Q_1 yields:

$$Q_1 = r_1 + R_e(1-p) - \tau_1\mu + $$

$$-\frac{1}{h}\sqrt{\begin{array}{l} h^2(R_e(2\tau_1\mu(1-p) + pR_e(1-p)) + \tau_1\mu(2r_1 + \tau_1\mu)) \\ +2h\mu(c_1(R_e(1-p) + r_1 - \mu\tau_1) + K_1 + pK_2 + E(BO)(pc_2 + \pi)) \end{array}}$$

REFERENCES

Barbarosoglu, G., and Y. Arda. 2004. A two-stage stochastic programming framework for transportation planning in disaster response. *J Oper Res Soc* 55(1):43–53.

Barbarosoglu, G., L. Ozdamar, and A. Cevik. 2002. An interactive approach for hierarchical analysis of helicopter logistics in disaster relief operations. *Eur J Oper Res* 140(1): 118–133.

Fenton, G. 2003. Coordination in the Great Lakes. *Forced Migrat Rev* 18(5):23–24.

Humanitarian and War Project. 2005. Available online from Feinstein International Famine Center, Tufts University, hwproject.tufts.edu/publications/electronic/e_croo.html (accessed August 26, 2005).

International Federation of Red Cross and Red Crescent Societies. 1999. *World Disasters Report 1998: Focus on Ethics and Aid*.

International Federation of Red Cross and Red Crescent Societies. 2005. *World Disasters Report 2004: Focus on Community Resilience*.

Minner, S., and O. von-Guericke. 2003. Multiple-supplier inventory models in supply chain management: A review. *Int J Prod Econ* 81–82:265–279.

Moinzadeh, K., and S. Nahmias. 1988. A continuous review model for an inventory system with two supply modes. *Manag Sci* 34(6):761–773.

Oh, S., and A. Haghani. 1996. Formulation and solution of a multi-commodity, multi-modal network flow model for disaster relief operations. *Transport Res* 30(3):231–250.

Oh, S., and A. Haghani. 1997. Testing and evaluation of a multi-commodity multi-modal network flow model for disaster relief management. *J Adv Transport* 31(3):249–282.

Ozdamar, L., E. Ekinci, and B. Kucukyazici. 2004. Emergency logistics planning in natural disasters. *Ann Oper Res* 129(1–4):217–245.

Sakakibara H., Y. Kajitani, and N. Okada. 2004. Road network robustness for avoiding functional isolation in disasters. *J Transport Eng–ASCE* 130(5):560–567.

Thomas, A. 2003. *Humanitarian logistics: Enabling disaster response*. San Francisco: Fritz Institute.

Thomas, M. U. 2002. Supply chain reliability for contingency operations, in *Proc Ann Reliab Maintain Symp*, 61–67.

UNICEF. 2003. *Programme Policy and Procedures Manual: Programme Operations*. UNICEF: New York.

UNICEF. n.d. UNICEF Operations in Southern Sudan. Monthly Reports. www.unsudanig.org/publications/ols/ (accessed August 26, 2005).

United States Agency for International Development, Office of U.S. Foreign Disaster Assistance (OFDA/AID). 2005. *Field Operations Guide for Disaster Assessments and Response*.

United States Committee for Refugees and Immigrants. 2004. World Refugee Survey 2004 Country Report: Sudan. http://www.refugees.org/countryreports.aspx?area=investigat e&subm=19&ssm=29&cid=170 (accessed August 26, 2005).

United States Holocaust Memorial Museum. 2005. Sudan: South/NUBA MTNS. www.ushmm.org/conscience/alert/sudan/ (accessed August 26, 2005).

7 Seeing Inventory as a Queue

Gerry Frizelle
University of Cambridge
Cambridge, United Kingdom

CONTENTS

7.1 INTRODUCTION

There have to be good reasons to look at the control of inventories from a queuing perspective, as the two subjects seem to have little in common. Such a viewpoint must offer insights and measures that are otherwise unavailable. This chapter explains the ideas and illustrates the benefits that can accrue from adopting such a standpoint.

Inventory control is concerned with how to order and monitor inventory levels to provide a company's customers with a level of service that, at the very least, meets their requirements at a competitive cost. Thus the emphasis is on predictability and how best to attain it. Classical queuing theory considers random processes and is couched in probabilistic terms. Its aim is to build mathematical models of situations where queues can occur. Usually the goal is to try to minimize the amount of time lost waiting in queues. Thus the two topics occupy opposite extremes of a spectrum of uncertainty ranging from the deterministic to the wholly random.

While both employ mathematical models, basic inventory control theory is relatively straightforward and intuitive; although the fundamental ideas have been embellished over time. Queuing theory, on the other hand, requires a degree of mathematical sophistication, even when dealing with the most basic models. It suffers from two further handicaps. The first is that many real-world situations cannot be precisely handled, in the sense that they lead to useful formulae.[1] Often only average behavior can be described. Unfortunately, excessive stocking costs are commonly incurred when the system's behavior is anything but average. The second difficulty is that models can change quite dramatically with small changes in the structure of the system being modeled. Examples of such changes are in the number of servers employed (a generic term for the mechanism that processes the entities forming the queue), how they are linked, the pattern of arrivals, the pattern of service times (i.e., how long it takes to process each entity), and so on.

What then does a queuing standpoint bring and how can the shortcomings of the approach be surmounted? The answer to the first question is that it provides a dynamic view of inventory. More precisely it permits us to compare what was planned with what actually happened. Indeed we can go further and then examine the causes for these deviations from the plan and rank them by their severity. No counterpart of such an approach seems to be available in traditional inventory control theory. The answer to the second question comes from the use of yet another theoretical tool—information theory. This reveals an entirely unexpected link into a specific form of queuing behavior. It indicates that inventory can, in some ways, be regarded as a special type of queue and is typical of the unsuspected relationships that mathematics sometimes uncovers.

This chapter, therefore, is structured in the following way. Sections 7.2, 7.3, and 7.4 give some background on the three areas just mentioned starting with the briefest summary of those parts of traditional stock control theory that are relevant to the discussion. The ideas will probably be familiar to readers. Then comes a short introduction to the simplest models in queuing theory as, fortunately, these are all that we will require. The third section sets out some of the concepts that underpin information theory. More depth is provided here as the material is probably less well known.

Section 7.5 attempts to bring the ideas together and explains the all-important link between information theory and queues along with the consequences. After that, Section 7.6 is devoted to an illustrative example to demonstrate, among other things, how the data are analyzed. The aim is to provide, in quantitative terms, an understanding of what drives the dynamics of a particular inventory system.

The case study in Section 7.7 sets out to show how these ideas can and have been used in practice. It records the findings that were obtained on the workings of a complex network of supply chains. It illustrates how viewing inventories as dynamical queues can identify and prioritize problem areas in a way that is not possible simply by applying standard inventory control techniques.

Throughout, mathematics will be kept to a minimum with only key formulae presented and explained. Our goal is to provide understanding and insight, not proofs. However, notes are provided at the end of the chapter. Here more formal and technical explanations can be found about topics appearing in the main body of the chapter, but where such detail would be a distraction. These will be indicated by a superscript in the text. The list of references at the end is intended primarily for readers who would like relevant background. So, for example, they will provide proofs of the relevant theorems behind the standard results quoted in the chapter. The list is not meant to be exhaustive.

7.2   BASIC INVENTORY CONTROL

This section summarizes those parts of inventory control theory that will be needed in the remainder of the chapter. The focus is only on so-called independent demand inventory control (the words *inventory* and *stock* will be used interchangeably; Axsäter 2006). Here we assume the existence of some form of warehouse where inventory accumulates. It will contain a variety of different products; these are referred to as stock-keeping units (SKUs). The demand for any one product (SKU) is independent of the demand for any other. Products are ordered from a supplier—this could be an internal supplier—and are shipped to a customer. There can, of course, be several suppliers and several customers involved. Independent demand means that we need to only consult the demand history for the particular SKU in which we are interested.

The goal of inventory control is to minimize the total cost of inventory. It balances the costs of holding stock with the cost of ordering, the so-called economic order quantity, and the cost of lost business through being unable to supply. One consequence of independent demand is that the total inventory costs will simply be the sum of the costs of holding individual items.

The inventory that is held is sometimes classified into two categories, active stock and buffer stock. This is shown in Figure 7.1, as the so-called saw-tooth diagram, for a single SKU. Active stock is the inventory that is assumed to be required; buffer stock is inventory held just in case. In other words, with total knowledge of the future demand pattern and assured deliveries, a buffer stock is not required. However, uncertainty can arise in at least three ways: changes to the level of demand, suppliers not meeting delivery promises, and failing to deliver the quantity ordered.

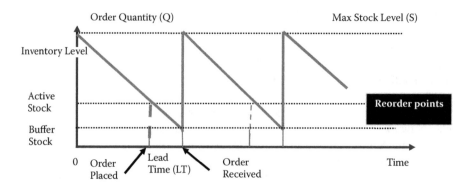

FIGURE 7.1 Idealized (saw-tooth) inventory diagram showing stock level over time.

7.2.1 ACTIVE STOCK

Figure 7.1 shows the opening active stock just equal to the known level of demand. Replenishment supplies are available, provided a new order is placed within a specified lead time (*LT*). This is when the inventory level reaches the reorder point. Reaching the reorder point triggers the new order. The amount ordered (the reorder quantity, *Q*) just equals the known demand for the next period, and, on receipt, brings the level back up to the maximum stock (*S*). The order arrives at the point at which the existing stock is equal to the buffer stock.

Of course, the real world is not so well ordered; the saw-tooth will not have the regular pattern of Figure 7.1. The demand slope can vary, steeper when the demand is increasing and vice versa. Deliveries can be late or early and so on. Therefore, in practice, one of two broad policies is usually followed. The first is always to order a fixed quantity. Here *Q* is fixed so that whenever the reorder point (*s*) is reached, the quantity (*Q*) is reordered. A variant of this is to fix *S*, the maximum stock (instead of *Q*). Now whenever the reorder point is reached, the amount reordered always brings the stock up to the maximum value. The second approach is to work to a fixed reorder frequency, for example, we might decide to order every week. In that case the amount ordered will depend on the inventory level at the end of the week; the reorder point is no longer relevant. Here too the rule is usually order enough to restore the level to some fixed maximum value.

The important point is that the active stock reflects what is known in advance; management have full knowledge of the demand and delivery patterns, even where these may be varying.

7.2.2 BUFFER STOCK

The buffer stock is the inventory level between the lower dotted horizontal line, in Figure 7.1, and the x-axis. Its role is to cover unexpected events. Where these increase either in severity or frequency, the buffer stock likewise needs to be increased. Were events to be totally predictable, no buffer would be required. In that case, the next order would be delivered at the point where the stock ran out. It also follows that holding a buffer stock adds to the total costs. Therefore, one way to make the operation more cost-effective is to take steps that allow the buffer stock to be reduced.

7.2.3 CONTROL PARAMETERS AND PERFORMANCE MEASURES

It is worth reiterating that management chooses the key control parameters: the reordering mechanism (reorder point or reorder frequency), how much to reorder, when to reorder, and what level of buffer stock to hold.

The principal measures of the system's performance include how well the company meets customers' demands, average levels of inventory held, the money tied up in inventory, and the number of stockouts. Most systems will probably include other metrics such as the performance of the suppliers delivering to the warehouse.

7.3 QUEUING THEORY

The basic ideas behind queuing theory are simple and are summarized in Figure 7.2. Entities arrive to be serviced at a "server" and, when the service is complete, they leave. If the server is busy when an entity arrives, it is forced to wait and a queue is formed. The entities arrive at a certain rate, the *arrival rate,* and are serviced at a certain rate, the *service rate.* Queues are so common that it is scarcely necessary to give examples; most people have experienced queuing at checkouts in supermarkets or waiting to get through airports.

7.3.1 GENERAL RESULTS FOR QUEUING THEORY

Queuing theory attempts to answer a number of important questions: When is a queuing system stable, in a statistical sense? What is the average waiting time for entities in the queue? What is the average service time? How long might a queue become? The answers help in designing facilities that will have adequate capacity.

Surprisingly, however, there are relatively few general results available. Two important ones are given here. The first derives the general condition for stability in a queuing system. It says that the arrival rate of entities at the server must be less than the service rate—with one important exception discussed in Section 7.3.2. The result is written mathematically in Equation 7.1:

$$\rho = \frac{\lambda}{\mu} < 1 \tag{7.1}$$

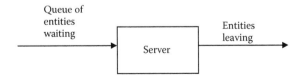

FIGURE 7.2 The basic input–output structure of a simple queue.

where ρ is called the *traffic intensity*, λ is the arrival rate, and μ is the service rate. When the utilization factor is greater than one, we have instability with queues that keep on growing, sometimes called explosive queues. The second important result is Little's law. This relates the average number of entities in the system to the average time in the system and the arrival rate. The formula is given in Equation 7.2:

$$\bar{N} = \lambda T$$

$$(7.2)$$

where $\bar{N}$ is the average number in the system, λ is the average arrival rate, and T is the average time spent in the system. Thus knowing, for example, the queue length and the arrival rate, the average time for entities to get through the system can be calculated.

7.3.2 Types of Queues

The simplest form of queue comprises a single server, entities arriving one at a time, being serviced on a first come, first served basis and leaving immediately after the service has been completed. The arrival time is random and the service times are also randomly distributed.[2] This type of queue is given the following three-part description; it is called an *M/M/1* queue. *M* stands for Markovian (named after Markov, one of the founders of the theory of random processes). The first letter describes the arrival pattern and the second the distribution of the service time; the third part gives the number of servers. In this case we have a particularly simple formula, shown as Equation 7.3, giving the probability that there are k items in the system.

$$p_k = (1-\rho)\rho^k$$

$$(7.3)$$

where p_k is the probability of having k entities in the system, and ρ is the traffic intensity. The two points about this equation are that shorter queues are more likely than long queues, and that the chance of the server being idle is $(1-\rho)$. Thus if the utilization factor is 0.75, then, not surprisingly, the server will be idle for around 25 percent of the time. Expressions for the other measures, such as average time in the system, can then be written down.

There is one other type of queue to mention. This is the *D/D/1* queue, where *D* stands for deterministic. In other words, the arrival times of entities are precisely known, as are the service times. This is the one queue where ρ can be equal to one. Here we are able to time the arrivals to come at exactly the point where the service on the preceding entity has been completed.

7.3.3 The Impact of Randomness

The *D/D/1* queue and the *M/M/1* queue sit at opposite ends of a spectrum of increasing randomness.[3] The impact of increasing randomness can be illustrated by looking at the number of entities in the system. In the case of the *D/D/1* queue, if the next entity arrives when the previous one has completed service, then there is never more than one in the system. Indeed if the traffic intensity is 0.75, then there need only be an average number of 0.75 entities. In this case an entity completes, then the server sits idle for 25 percent of the cycle time, at which point the next entity arrives to be serviced. However, in the case of the *M/M/*1 queue, the average number in the system is three (the formula for the average number in the system is given by $\rho/(1-\rho)$). Thus the price for increasingly random behavior is longer queues.

7.3.4   VIEWING INVENTORY AS A QUEUE

Before leaving the topic, we need to introduce the idea of a *virtual server.* This is when there is no actual service involved but merely a delay. In such a case the assumption is that "service" begins when the entity arrives at the queue and is complete at the point where the entity leaves the queue. If we combine this idea with that of the *D/D/*1 queue, we have an alternative description of the active stock mechanism shown in Figure 7.1, albeit one where "deliveries" are of single items only. However, the queuing model can be modified to handle bulk arrivals, equivalent to delivering a reorder quantity, *Q*.

So far we gain little from taking a queuing perspective, apart from two insights. The first is that we should not try to run the system flat out, where $\rho = 1$, unless we were to exercise total control, which is, of course, impossible. The second is that we pay for loss of control in terms of greater inventory; hardly a surprising result. The other measures furnished by the queuing theory are nothing more than we calculate anyway in inventory control. For example, the average number in the system, $\bar{N}$, is simply the average inventory level. What is more, this number can be recorded directly in a warehouse, not derived from a mathematical formula of only limited applicability.

Something is missing. We need to develop a measure that can be calculated by directly observing the inventory (queue) sitting on a shelf while providing insights not available from the existing inventory control theory. In addition we want to employ some of the ideas and structures from queuing theory but without the shortcomings highlighted earlier, such as having to build a mathematical model for each type of queue we encounter. We would then have a more empirical and hence more robust approach.

7.4   INFORMATION THEORY AND ENTROPY

The missing link is provided by information theory. This section spends a little time explaining the ideas, as, in contrast to queuing theory, some of the concepts are quite subtle, but the mathematics are relatively straightforward.

Information theory is largely the creation of one man, Claude Shannon, developed while he was working at Bell Laboratories at the Massachusetts Institute of Technology. His interest was in the mechanics of sending and receiving messages. He envisaged a sender linked in some way to a receiver, possibly through a wireless connection. This link is called the *channel of communication*, or *channel*. He realized that to send a message remotely, it had to be translated into a form that could be transmitted. For example, the earliest wireless messages were sent in Morse code. This translates letters of the alphabet into a series of dots and dashes (*characters*). He described the activity as *coding* the message. The message would then need to be decoded at the other end by the receiver. It meant that the sender and receiver had to agree in advance on a coding scheme.

Shannon (Shannon and Weaver 1999) asked himself what corresponded to an efficient code, that is, one that was flexible enough to enable any message to be encoded but was as short as possible (*data compression*); the shorter the length of the encoded message, the more quickly it could be sent. In addition, frequently used letters of the alphabet should have shorter codes (*codewords*) than those used infrequently. For instance, although Morse code is not a particularly efficient code, the letter E is coded as a single dot.

Shannon was also interested in how quickly a message could be sent (the *transmission rate*). This turns out to be the capacity of the channel; the so-called channel *bandwidth* (Cover and Thomas 2006). Linked to that was the important issue of noise—roughly speaking, characters received that have nothing to do with the message. Clearly the presence of noise will reduce the effective capacity of the channel. One of Shannon's extraordinary discoveries was that it is possible to transmit nearly noiseless messages provided that the channel capacity is not exceeded.

7.4.1 WHAT IS INFORMATION?

Information (also known as self-information), as defined by Shannon, is a measure of the content of information associated with an event. It is defined, in mathematical terms, as

$$I(X) = \log_2\left(\frac{1}{p(X)}\right) \tag{7.4}$$

Here I is the information, X is the event, and $p(X)$ is the probability of the event happening. Its unit of measure is a bit. This rather extraordinary definition needs some explaining. The first question is why is there a probability involved. This is because information is something new, that is, news. Thus in the case of the sender and receiver, it is data the sender has and the receiver wishes to have. So from the receiver's point of view, the content is news (otherwise why send the message?). The next question is why is the reciprocal of the probability used. The answer, that people sometimes find difficult to understand, is that the rarer the event, the more information it carries. We can see this by writing a sentence first omitting the vowels, then the consonants. Thus taking the first phrase of the preceding sentence and leaving out the vowels gives "W cn s ths by wrtng sntnc"; then leaving out the consonants gives "e a ee i ii a eee." We might make a reasonable guess at what the first version of the phrase was, but the second is unintelligible. This is because consonants occur less frequently than vowels (on average) and thus carry more information.

The third aspect of this formula is the introduction of a logarithm. This is a little harder to explain. Basically a message is a string of characters (letters, numbers, etc.). Each character will contain a certain amount of information. Taking logarithms is a way of "counting" the number of characters in the string, that is, the information content. Taking logarithms to the base 2 reflects the fact that we are in a digital age where we deal, ultimately, with only with ones and zeros. The definition of a bit only applies when we employ logarithms to the base of 2.

7.4.2 ENTROPY

The idea of counting will become clearer when the concept of entropy is explained. Entropy (or more correctly entropy rate) is the lower limit to which a message can be compressed. It is also a measure of the *average* random uncertainty in a message. Entropy, as defined by Shannon, is given by the following formula:

$$H(X) = -\sum_N p(x)\log_2 p(x) = \sum_N p(x)I(x) \tag{7.5}$$

Here H is the symbol for entropy, X is a random variable describing an event, and $p(x)$ is the probability associated with the occurrence of X (formally of X taking the value x). N is the number of random variables involved.

Equation 7.5 will look strange to anyone who has encountered entropy in physics or chemistry. However, it is the same concept, just formulated in a slightly different fashion. Entropy (and information theory) turns out to be an idea of extraordinary power and generality (Cover and Thomas 2006). Moreover it has proved to be more important and useful than information, as given in Equation 7.4. Unfortunately, exploring this fascinating topic further would take us too far away from the subject of the chapter.[4]

The first version of Equation 7.5 is the one most commonly used. The second version shows the link to the information, in Equation 7.4.[5] Notice one important characteristic of Equation 7.5: it does not depend explicitly on the choice of random variable. Thus, in theory at least, it can be applied to any random variable, providing that the random variable is properly defined.

7.4.3 DATA COMPRESSION AND ENTROPY

It will now be shown how Equation 7.5 is applied, how entropy is linked to data compression, and how taking logarithms has the effect of "counting the information content."

Suppose that we plan to run a warehouse that will contain eight distinct SKUs. Our first task is to devise a labeling system. Our main (rather improbable) criteria are that the label should be a binary string and be as short as possible. Equation 7.5 can tell us how long the label will need to be. However, as we are not yet in business, we have no idea which units will sell well and which will not. Therefore, we must assume that all are equally likely to sell (so-called equiprobability). Thus we assign a 1/8 chance that a particular SKU will be ordered when a sale is made. Putting these probabilities into the entropy formula (Equation 7.5) gives Equation 7.6.

$$H(\text{labeling SKUs}) = -\sum_{i=1}^{8} p_i(\text{single SKU})\log_2 p_i(\text{single SKU})$$

$$H(\text{labeling SKUs}) = -\sum_{i=1}^{8} \frac{1}{8}\log_2\frac{1}{8} = \log 8 = 3 \tag{7.6}$$

(recall that $-\log p = \log p^{-1} = \log (1/p)$). This calculation tells us that we will require a label of, at least, length three to give each item a unique identification. It does not say that every label will be of length three, but that there will be at least one of length three.

To see this, assume that we simply call the SKUs zero, one, two, three, and so on but in binary. The first three products will be labeled 0, 1, 10. The eighth will be 111, a label of length three. Thus, the entropy formula has linked the number of SKUs—the variety of stock items we require—to the (minimum) label lengths.[6]

This also explains why the use of logarithms introduces a way of counting. Recall that 8 can be written as 2^3. When we take the logarithm (to the base 2) of this number we get $3 \log_2 2$. This equals 3, the minimum label length and says that the information content is 3 bits. Had we had only four SKUs then the information content (formally, the chance of selling a particular SKU) would have been two; you can see this by inserting 1/4 into Equation 7.5.

One important final point is that the entropy takes its maximum value (where the number of outcomes are finite) when all outcomes are equiprobable, as in this example. Equiprobability is another way of saying we have no prior knowledge—in this case of the level of future sales. Equation 7.5 takes the simple form shown in Equation 7.7, where N is again the number of random variables. In our example, this is the number of SKUs we will be storing in the warehouse.

$$H_{max}(S) = \log_2 N \tag{7.7}$$

Suppose now that the warehouse has been running for some time and we want to introduce a computer-based inventory system. We wish to minimize the amount of memory space used, because every time we have a stock transaction we need to record the label of the SKU involved. Might we be able to use the additional data we have on sales history to reduce the call on memory? We can do this by assigning the shorter labels to the SKUs most in demand. Hence we would label the most popular SKU "0," the second most popular SKU "1," and so on. Thus the most frequently recorded transactions, making the greatest demand on memory, will have the shortest records and vice versa.

7.4.4 ENTROPY AND COMPLEXITY

Entropy has yet another useful property. It provides a measure for one form of complexity.[7] This type of complexity may be thought of as being generated by a system that can evolve in a number of different ways. These different ways are called the possible trajectories the system can follow over time. Each possible trajectory has an associated probability, reflecting the chance of the system actually selecting it. The entropy is therefore the amount of information, averaged over the trajectories, required by the observer. Clearly, the more trajectories available to the system, the more complex it is.

7.5   BRINGING THE STRANDS TOGETHER: ENTROPY, QUEUES, AND COMPLEXITY

All of this is somewhat peripheral to inventory control. It still needs to be established how the ideas developed so far are linked together. The first step is to model the process of taking observations; in our case, the movement of inventory in and out of a warehouse.

7.5.1   MODELING THE OBSERVATION PROCESS AND SECRETIVENESS

The process of observation is shown in Figure 7.3. This diagram suggests that there may be parallels between taking observations and the passing of messages. We might think, rather fancifully, of the system "sending messages" to the observer, which he or she then records. There is a major difference, however: the system's goals differ from those of the observer. The former is to offer a level of customer service, the latter to record accurately what is happening. One consequence is that the observation process must actively avoid impacting on the system as it might interfere with the latter achieving its goals. For example, if the recording mechanism itself impacted the accuracy of the resultant record, then deliveries could be compromised, adversely affecting the level of service. Such a situation used to occur when inventory movements were recorded manually. Stock records sometimes might only be updated at the end of a working day, or even the end of a week. In the interim, dispatch promises were made based on incorrect data.

This noninvolvement is described as treating the system as though it were *secretive* (Frizelle and Suhov 2008). In practical terms it means that, when we come to analyze how the inventory system is performing, we have to rely solely on our record. So, for example, we cannot interrogate the system—except possibly in extreme circumstances.

However, secretiveness is also an issue even when setting up the system. Suppose we decide to create an inventory control package to sell as a commercial product. We need to specify our new offering. Two obvious fundamental parameters will be how many SKUs our product can handle and how big the stock records will need to be. Might we have customers that have millions of items in stock, for instance? Clearly what we need in each case is a number that is big enough to guarantee that we can accommodate whatever requirements our prospective customers demand of us. How do we calculate this number? The proposed system itself cannot tell us. We might try to find what existing competitive systems do, we might try to discuss the matter with potential customers, we might survey the literature, or do all of these things and more. However, the fact is that such a number can never be calculated (this is a fascinating

FIGURE 7.3 The stages of an observation and recording process..

topic in its own right; interested readers are directed to Cover and Thomas [2006] or Hofstadter [1979]). All we can do is make well- or ill-informed guesses.

7.5.2 OBSERVATIONS AND MEASUREMENT

Returning to Figure 7.3, once we have created our record, we will want to use it to carry out some form of measurement. The goal will be to assess how well the inventory control system is performing. That generates more questions: What precisely do we observe? What measure do we use? Do we need more than one measure?

Now there is a surprising relationship that allows us to address the first two questions. It turns out that our inability to ever provide definitive answers to such questions as "how many SKUs do we allow for?" when designing the system, holds the key. We have to conclude that, in general, we are never going to be able to determine an upper limit in the total certainty that it will never be breached—although it may be possible in specific instances. Stating this in mathematical terms, we must work on the assumption that the systems we are dealing with can have an infinite number of states.[8] We must be careful about what we are saying; it is that we are never able, in general, to determine an upper limit, so we have to treat systems as if no such limit exists.

However, as the reader will probably have guessed, we will want to use entropy as our measure. That creates a problem, as Equation 7.7 now takes the value of infinity. It raises practical difficulties if we want to use it as a measure, particularly when comparing two systems.

The question is then, is it possible to have a system with an infinite number of states but finite entropy? The answer is yes. We can find a mathematical condition that puts an upper limit on the entropy per state and ensures that the overall entropy is finite, even though the number of states is not.[9] We can state the condition for the entropy to be maximized but remain finite; the probabilities must follow the so-called geometric distribution.

Now comes the surprising bit: This distribution has exactly the same structure as the formula for the $M/M/1$ queue. The only difference is that the probabilities are replaced by the traffic intensity. In other words, for the input–output system shown in Figure 7.2, the entropy of the system is linked to the behavior of the queues. This completes the jigsaw for it suggests that when looking at input–output systems, we need to observe queuing behavior, and when we take an entropy measurement we measure queuing entropy. Moreover, when we compare two systems, we do so on the basis of the dynamics of their respective queues. It tells us, among other things, which of the two exhibits the most complex behavior.

There is more. When the queue behavior can be modeled as $M/M/1$, its entropy is at a maximum for the specific value of the traffic intensity. This is what was meant when it was stated at the start of the chapter that queues (specifically $M/M/1$ queues) sit at one end of a spectrum of uncertainty. Equally, if the system was to be wholly deterministic, the $D/D/1$ queue, then the entropy will be zero—the other end of the spectrum. It says, in effect, that we can model a real-world queue as being composed of two elements: a deterministic bit and a random bit.

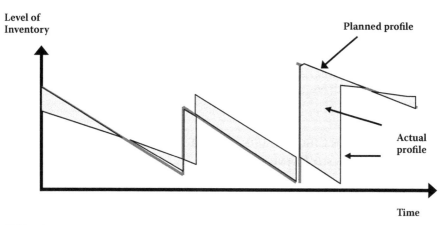

FIGURE 7.4 The planned inventory profile against the actual one.

7.5.3 APPLYING THE IDEAS TO INVENTORIES

The notion of active stock was introduced at the start of the chapter, as represented by the saw-tooth in Figure 7.1. As pointed out, everything is known (deterministic): the demand, the delivery time, the delivery quantity and so on. Thus active stock represents variety without probability. We can go a stage further and say that achieving the stock profile represented by the active stock is the immediate goal of the stock control system

In practice, of course, we can never be wholly certain. If we look at the actual inventory profile of a SKU over time, it will be the conflation of the planned and unplanned. However, if we know what was meant to happen, then we can separate the planned from the unplanned.

The idea is shown in Figure 7.4. The light gray is the planned profile—the goal the system is trying to achieve. It has been drawn in an irregular fashion to emphasize the fact that the shape of the profile is irrelevant. This line has to be "constructed," as it is unlikely to exist in any inventory records. It is done by looking back at planned orders and deliveries. The dark gray line is the actual profile and this can be drawn directly from the inventory records. The shaded area is the variation between the two; the unexpected element giving rise to the entropy.

7.6 ANALYSIS

Two key pieces of analysis are carried out. The first is to assign causes to the variations between the planned and actual performance. There is, of course, no definitive list of such causes. However, typical examples include customers changing their minds (e.g., canceling orders, modifying orders, rescheduling orders, replacing orders), the inability to dispatch to customers through having insufficient stock on hand, late delivery by suppliers, delivery shortages, and erroneous deliveries from suppliers. The list is determined by the specific nature of the stocking operation, what that particular management wants to know, and what data are available.

The second key piece of analysis is making comparisons. These can be between the performance of individual SKUs, between suppliers, or between customers. Other useful data include whether incomplete deliveries are more of a problem than late deliveries. We might assume that if we ordered one hundred of a particular SKU, it is preferable to have, say, thirty delivered on the day required with the balance later, than having all one hundred delivered late. However, it may not prove to be so in practice.

7.6.1 CALCULATING THE VARIATION OF THE ACTUAL FROM PLANNED INVENTORY PROFILE

Table 7.1 demonstrates how the calculations behind Figure 7.4 are carried out. The table shows deliveries into and out of a warehouse for a single SKU. Under "Planned," columns A and B record the quantities planned to be delivered from the supplier and the dispatches required by the customer. The resulting planned inventory levels, at the end of each day, are given in column C along with the opening stock. Under "Actual," columns D, E, and F show the actual performance under the same three headings. The inventory variation is recorded in column G.

It will be clear from this example that poor delivery to the customer is being caused by poor delivery from the supplier. Thus, the first order that was supposed to be delivered on Monday does not come until Tuesday (see columns A and D). On Wednesday the supplier only delivers ten instead of the thirty required. Thursday's order is delivered a day late and then not in full, as twenty of the quantity delivered form part of Wednesday's order. This inability to supply causes violent fluctuations in the stock levels, with a stockout on Thursday. However, that is not the whole story. On Monday there was sufficient inventory in the warehouse to make a full delivery of eighty (column F, opening stock). However, the customer changed his mind and asked for only fifty, with fifty the following day and the balance on Wednesday

---

TABLE 7.1
Calculations Behind Figure 7.4

| Day | Planned | | | Actual | | | |
	Delivery from Supplier (A)	Dispatch to Customer (B)	Inventory Level (C)	Delivery from Supplier (D)	Dispatch to Customer (E)	Inventory Level (F)	Difference (G) = (F) – (C)
Opening inventory			100			100	
Monday	50	80	70	—	50	50	−20
Tuesday	—	50	20	50	50	50	+30
Wednesday	30	—	50	10	30	30	−20
Thursday	50	40	60	—	30	—	−60
Friday	—	50	10	50	—	50	+40

TABLE 7.2
Detailed Representation of the Results Shown in Table 7.1

	Supplier Performance			Dispatch Performance			
Day	Planned Delivery (A)	Actual Delivery (B)	Difference (C) = (B) – (A)	Planned Dispatch (D)	Actual Dispatch (E)	Difference (F) = (E) – (D)	Principal Causes (G)
Opening inventory		100			100		
Monday	50	—	–50	80	50	–30	Customer changes schedule
Tuesday	—	50	50	50	50	0	Customer changes schedule
Wednesday	30	10	–20	—	30	30	Customer changes schedule
Thursday	50	—	–50	40	30	–10	Late delivery by supplier
Friday	—	50	50	50	—	–50	Late delivery by supplier

(column E). This was because the customer was unable to take the whole eighty in a single delivery.

A second way to carry out the analysis is shown in Table 7.2. Here the supplier and the warehouse performance are separated: the former is given in columns A, B, and C; the latter in columns D, E, and F. The causes, summarized in the previous paragraph, are recorded in column G. Notice that no mention has been made here about the supplier's late deliveries on Monday, Tuesday, or Wednesday. This is because a key role of a warehouse is to provide a reliable dispatch service to the customer and is one of the principal reasons for holding inventory. Thus while poor service from the supplier was a problem, it is not the major cause for the pattern of dispatches on those three days.

7.6.2   ESTIMATING PROBABILITIES

The next stage in the analysis is to calculate the relative frequencies. So, for example, in Table 7.1 in column G, –20 appears twice while 40, 30, and –60 each appear once. Therefore, an estimate of the associated probabilities (based on a very small sample of five) are 0.40 for –20, and 0.20 each for 40, 30, and –60. Notice that we can do the same in column G in Table 7.2, giving a probability estimate for "Customer changes schedule" as 0.60 and a probability estimate for "Late delivery of supplier" as 0.40. Once the probabilities have been estimated the entropy can be calculated from Equation 7.5.

7.6.3   USING ENTROPY AS A MEASURE

We have still not answered one of the questions posed at the start of Section 7.5.2: What measure do we use? Or more to the point, why do we select entropy as the measure?

There are three practical reasons and one theoretical one. The first practical one is that entropy aggregates variation, in a way that probabilities do not; the latter always sum to unity. You can think of it as an indicator of the level of turbulence in a system—the system behaving in unexpected ways. The greater the tendency for this to happen, the more turbulent (complex) the process becomes. In terms of Figure 7.4, the two lines will progressively diverge from one another as turbulence increases.

The second reason is that it is possible to allocate levels of turbulence to their causes.[10] Thus, for example, we can differentiate between causes about which we can do something and those we cannot.

The third reason is that we can make comparisons, thanks to the aggregating properties of entropies. Thus we can compare supplier performance by carrying out calculations based on columns A, B, and C of Table 7.2, for different suppliers. Indeed we can go further and look at how much disruption a particular event causes when it happens as against how much of a problem it causes overall. For example, a particular supplier may deliver late and cause considerable inconvenience to a number of customers—very disruptive when it happens. However, it is rare for that supplier not to be on time so, overall, it is not a major issue.

The theoretical reason is that Equation 7.5 can be applied wherever a random variable can be defined and probabilities assigned to the various outcomes. This has the great advantage that any type of queue can be studied. It frees us from having to build a new model for every specific queue we encounter. This is a powerful advantage because we can make comparisons between very different stock profiles within the same warehouse, for example.

There are drawbacks in using the measure. The first and most obvious one is the very strangeness of the measure, not to mention its mathematical structure. Managers feel uncomfortable about being asked to trust a measure that they do not fully understand. Experience has shown that trust grows when it highlights matters with which the management is already familiar and when it is used in conjunction with more traditional metrics.

A second drawback is the very versatility of entropy as a measure. As we have just seen, entropy values can be calculated irrespective of the nature of the variable under consideration. This is also a weakness when comparing systems. The measure itself gives no guidance as to the validity of the comparison being made. We could just as easily compare chalk with cheese as chalk with chalk. We need some additional model to ensure that this does not happen. That is another reason why the link between entropy and queues is so important; it leads us always to compare queuing behavior.

7.7   CASE STUDY

7.7.1   BACKGROUND

The case study describes a project carried out in a large distribution center (DC). The project aimed to provide some insight into the performance of the supply chains feeding into and out from the DC. Located near Birmingham in the United Kingdom, this extremely complex operation involved a large number of suppliers, haulers, and end customers, as well as having a sizeable internal operation (Robinson 2008).

This exercise focused on three factories delivering nine products—SKUs—into the DC, which, in turn, dispatched them to three end customers. In total twenty-eight chains were involved (a chain is considered to be a supplier, a customer, and a product). The DC plays the role of customer to the factories and supplier to the end customers.

The advantages of looking at the performance of the various inventories within the DC is that this represents the one point where all the chains come together. It is also where the data can be found. Therefore the approach provides a consistent way to compare the performance of a disparate group of players.

7.7.2 DATA COLLECTION

Data collection was a three-stage process. The first stage involved accessing actual inventory records for each of the SKUs held in the DC, to obtain a historical plot of the actual inventory levels. These data were, conveniently, event based (in that a record was made every time the inventory level changed) and stored electronically.

Stage two involved a reconstruction of the planned inventory levels. These data came typically from orders and schedules placed on and by the distribution depot. Such orders indicate the dates and quantities required. In total, seven data sources were consulted.

The final stage was more problematical, assigning causes to the variations observed. One practical problem was that, with no one person or company in overall control, it was not always easy for the project leader to find and obtain permission to use the data. A second was that records were dispersed among the various parties involved; so that linking cause to effect was not always easy.

7.7.3 DATA ANALYSIS

A major simplifying assumption was made. This was that each of the supply chains could be considered in isolation from the others, except where a stockout of a particular SKU caused a dispatch to be delayed to an end customer. Thus, for example, the variation between what should have been delivered to stock and what was for one SKU was independent of the variation between what should have been delivered to stock and what was for a different SKU.

Making valid comparisons was a concern. It helped that queues (inventory levels) were always being observed and always at the same location. It enforced a degree of uniformity on the exercise. Nonetheless practical difficulties regularly cropped up. One such difficulty is when the same SKU exists in different package sizes, for example, pallets, boxes, cartons, and so forth. Counting had to be consistent, so the unit chosen was the smallest one encountered. Thus if a pallet was delivered to a storage location but was then split into cartons for onward delivery, the unit of measure would be cartons.

7.7.4 RESULTS

Figure 7.5 gives the key findings. It shows the major causes of disturbance in the network of queues under consideration. The six causes of inventory fluctuations

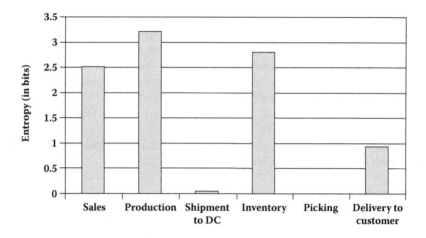

FIGURE 7.5 Causes of turbulence in the supply chains.

(virtual server states), shown on the horizontal axis, are explained in Table 7.3. The table shows that it was found possible to identify sources of variations. The first bar, "Sales," records the instances of end customers changing their minds between placing the order and the delivery. As can be seen, this is a major problem. However the "Production" bar is even higher. It suggests that the factories supplying the DC are not coping with the volatility of the customer demand. If anything they are making matters worse as this bar is significantly higher. It is surprising that all this turbulence appears to have been recovered by the time delivery is made into the DC, as reflected

TABLE 7.3

Description of How the Bar Chart in Figure 7.5 Was Constructed

Title Identifying the Bar	Planned Activity	Cause of Fluctuation
Sales	Planned orders from final customer to supplier (these orders bypassed the distribution warehouse)	Deviation of what was finally ordered from what was planned
Production	What each plant planned to produce	Deviation between what was planned and what was actually produced
Delivery to distribution center	Planned delivery to distribution center	Deviation from what was promised
Inventory	Expected inventory, as calculated from expected incoming and outgoing stock	Deviation of actual inventory figures from expected
Picking	Planned pick lists	Deviation of actual pick from plan
Customer delivery	Delivery, as agreed with the customer	Deviation of what was actually dispatched to customer from what was agreed to be delivered

by the size of the "Shipment to DC" bar. The reason for this was not altogether clear but seems merely to be a reflection on the haulers' performance—they delivered when they were asked to. In other words, the haulage stage from the factory to the DC has not aggravated the situation. What is clear is that by the time products come to be delivered to the end customer, turbulence is still present, although considerably attenuated from that seen further up the chains. It is concluded that holding intermediate inventory in a warehouse is not providing a total buffer against fluctuations generated further back in the operation.

Finally, the "Inventory" bar shows the impact of the various sources of turbulence on inventory performance, in terms of how far the actual level deviates from the planned level. Such fluctuations have to be covered by a buffer stock. Therefore one can—at least in theory—estimate the cost, in terms of excess stocking costs, of having a turbulent supply chain.

It was also possible to identify which factories were the problem suppliers, which customers created the greatest difficulties in terms of changing their minds about what to order, and which of the nine products proved the most difficult to manage through the chains. One particular factory was found to be having special problems.

The case study illustrates the benefits that can accrue when viewing inventory as a dynamic queue. It offers a way to assess the performance of a complex network of supply chains by providing a single measure at a key node in the network. The measure creates both an absolute performance metric (how closely the SKU achieves its goal of meeting the planned inventory profile) and a comparative measure (the relative performance of suppliers, customers, and other participants in the network). Where the investigation does not provide the complete answer, it at least highlights those locations that need to be examined in depth. Such an analysis would not be possible relying solely on the data provided by traditional inventory control systems.

7.8   CONCLUSIONS

We have attempted to demonstrate the additional benefits that accrue when we view inventories as a class of queue. The principal one is that it encourages us to consider the dynamical behavior of inventories. This is revealed in the inventory profile shown in Figure 7.4, which is part planned and part random. Moreover, the figure demonstrated how the two components can be separated; in other words, the total behavior is the sum of the predicted and the random elements. However, the viewpoint brings a second benefit. We say that a virtual server is involved. This assumes that service begins when an item goes into stock (joins the queue) and is complete when it leaves. It encourages us to think in terms of server states, that is, is the server behaving as it should or, if not, what is it doing. The important aspect of this rather odd standpoint is to highlight the link between server behavior and queue behavior. In rather more prosaic terms, we look to see what causes the queue to behave as it does and, more important, put a measure on the amount of turbulence that is created as a result.

In fact, this linkage of server to queue states allows the ideas to be applied to two forms of inventory not usually included in traditional stock control theory and, for that reason, not considered in the body of the chapter. However, they need a brief mention for completeness. The first is where a real server exists, for example, a resource in a factory (Frizelle and Suhov 2001). The queues involved are often referred to as work in process. The nature of the factory dynamics is then revealed by analyzing the server states, for example, is the resource working or in a state of breakdown. The second form of inventory is that found in industrial processes. Here we are dealing with gases, liquids, or solids. In mathematical terms, they are described by continuous variables. Inventory control is not an issue, apart from making sure that supplies are delivered when required. As the materials are usually stored in some form of container—tanks, silos, or bins—there is a physical upper limit on how much can be held. Indeed a process plant will have been designed with these capacities in mind.

We also described the benefits that come from using entropy as a performance metric. First, entropy is a measure of turbulence and is, more generally, a measure of one form of complexity. Second, it is independent of what is being measured. This means that we can compare similar systems, which in this context means the queuing behavior of different SKUs. It also releases us from first having to specify a queuing model to enable us to carry out measurements. Third, entropy provides an absolute measure, in the sense that no uncertainty means no entropy and maximum entropy is defined relative to the traffic intensity of the queue.

However, the real benefit comes from being able to apply the ideas in practice. Observing the queue dynamics in a large distribution depot provided insight into how a complex network of supply chains was performing. The analysis revealed likely causes of the problems and indicated where unnecessary costs might accrue through having to hold excessive buffers in the system. Such insights cannot be gleaned from ordinary stock control theory.

It was commented at the start of the chapter that adopting this perspective can only be justified if it provides some fresh viewpoints. The hope is that this has been demonstrated to be the case.

NOTES

1. The technical term for this is that no closed form of solution exists. It means that, in practice, one would probably have to use either approximations or some form of simulation to evaluate results.

2. These two statements have a precise mathematical definition for the simplest queues. The arrivals follow a Poisson distribution, whose probability mass function is given by $\frac{\lambda^k e^{-\lambda}}{k!}$, where k is the number of arrivals in a given time interval and λ is the average arrival rate. The service times are distributed as a negative exponential whose probability density function is given by $\mu e^{-\mu x}$. Here μ is the service rate and x is the time to complete a specific service.

3. As will become clearer further on in the chapter, the $D/D/1$ queue corresponds to a queue with zero entropy associated with it. The $M/M/1$ queue has maximum entropy, the value dependent on the traffic intensity.

4. There is an extraordinary link between entropy, complexity, and Turing machines. The latter is a sort of theoretical computer, named after its inventor, Alan Turing. They provide the conceptual underpinning of today's computers. For anyone interested, there is a very readable account in Cover and Thomas (2006) and a greater exploration of some of the ideas in Hofstadter (1979).

5. Readers familiar with statistics or probability theory will see that entropy is the *expected* value of the information. Notice how entropy aggregates all the contributing sources of information but gives more weight to likelier events and less to rarer ones, even though these latter carry more information.

6. You can try other coding schemes using binary numbers but you will always finish up with at least one SKU having a label of length three. For example, you might think that all possible permutations of 1 and 0 might be more efficient. Thus we would call our eight SKUs 0, 1, 00, 01, 10, 11, 000, and 001. Remember that these are not numbers but labels. Even so, we still need two labels of length three.

7. The form of complexity involved is called the Kolmogorov–Sinai complexity. Essentially it is about the limits of compressibility and defines the complexity of a string as the shortest computer program that can produce the string. For example, a string of totally random numbers will require a program of at least the length of the string—if a shorter program could be found, the numbers would not be random (even though there are programs that supposedly generate random numbers). This form of complexity is approximately equal to the entropy proposed by Shannon. This is part of the link alluded to in note 4.

8. By an infinite number of states, we mean countable infinite number. The important point here is that the states are discrete. Thus we can put them in a one-to-one correspondence with the natural numbers. An example of a countable set is a jobs-to-do list.

9. If this looks like mathematical jiggery-pokery, then think about $\sum_1^\infty \frac{1}{2^n}$, that is, adding 1/2 to 1/4 to 1/8 and so on. This sum never reaches 1 but we can prove that it "tends to 1," as the number of terms increases to infinity. Thus the number 1 is an upper bound on the series.

10. The equation for this is given by $H(A,B) = H(A) - \sum_{a \in A, b \in B} p(a,b) \log \frac{p(a,b)}{p(a)}$. Here A corresponds to the set of causes and B to the set of consequences, that is, the resulting queue turbulence of the causes. The term on the left-hand side of the equation is called the joint entropy and the double summation term is called the conditional entropy. The latter is the term that associates the causes of turbulence with their causes.

REFERENCES

Axsäter, S. 2006. *Inventory control.* Berlin: Springer.

Cover T. M., and J. Thomas. 2006. *The elements of information theory.* Hoboken, NJ: John Wiley & Sons.

Frizelle, G., and Y. Suhov. 2001. An entropic measurement of queuing behaviour in a class of manufacturing operations. *Proc Roy Soc A: Math Phys Eng Sci* 457(2001):1579–1601.

Frizelle, G., and Y. Suhov. 2008. The measurement of complexity in production and other systems. *Proc Roy Soc A: Math Phys Eng Sci* 464(2098):2649–2668.

Hofstadter, D. R. 1979. *Godel, Escher, Bach: An eternal golden braid.* New York: Basic Books.

Kleinrock, L., and R. Gail. 1996. *Queueing systems: Problems and solutions.* New York: Wiley.

Robinson, M. June 2008. *Complexity observed in supply chains. But can it be diagnosed?* Paper presented at Complexity in Business IET, London.

Shannon, C. E., and W. W. Weaver. 1999. *The mathematical theory of communications.* Urbana, IL: University of Illinois Press.

8 Fuzzy Inventory Models

Alfred L. Guiffrida
Kent State University
Kent, Ohio

CONTENTS

8.1 INTRODUCTION

The efficient management of inventory requires managers to make decisions on two fundamental questions: (1) how large should an inventory replenishment order be, and (2) when should an inventory replenishment order be placed. These two decisions are integral in the support of an overall inventory management program that typically has the objective of providing a desired level of customer service at a minimal cost. Beginning in 1913 with Harris's economic order quantity (EOQ) model, numerous mathematical inventory models have been published in the inventory literature to assist managers in determining the size and timing of an inventory replenishment order in support of this objective.

Mathematical inventory models can be broadly categorized into two groups: deterministic models and stochastic models. Deterministic inventory models were the first class of inventory models to evolve. In the years following the introduction of the Harris EOQ model, numerous variants and extensions of the EOQ model rapidly appeared in the literature. Deterministic inventory models typically contain few parameters, and all input data for these parameters are assumed to be known with complete certainty. The total cost function includes the costs associated with ordering, holding, and purchasing inventory. The optimal order quantity that minimizes the total cost function, then, can be easily found. Deterministic models also assume that the demand and lead-time characteristics of the inventory system are known with complete certainty, leading to a simple and straightforward mathematical structure for modeling demand during lead time.

Stochastic inventory models first appeared in the literature in the early 1950s. These models were formulated to capture the uncertainty and variability that exist in most real-world inventory situations but were assumed away in deterministic inventory models. As a result of incorporating uncertainty in demand and lead time, stochastic models require more model parameters than deterministic inventory models, and require that some input parameters be defined in probabilistic terms. The optimal solution of a stochastic inventory model generates a set of policy variables, which minimize an expected total cost function. Seminal papers by Arrow et al. (1951) and Dworetzky et al. (1952a, 1952b) established the foundation for stochastic inventory models and led to the introduction of shortage costs.

Stochastic inventory models assume that demand, lead time, or both factors can be described by random variables. The underlying mathematical structure for a stochastic inventory model is more complex than that of a deterministic model and requires a policy statement from management for the inventory system performance measure such as a service level or fill rate.

In practice, implementing either form of inventory model requires the acceptance of a set of assumptions that govern the model use, as well as the recognition that estimates must be made for model parameters such as demand rates, lead times, and inventory-related costs.

Obtaining realistic input values for the parameters of a mathematical inventory model is a challenging task. The analyst performing this task is often operating in a decision environment that is characterized by uncertainty, ambiguity, and the lack of perfect information on customer demand rates, manufacturing, and delivery lead times, in addition to inventory cost structures for ordering, holding, and shortages.

Fuzzy set theory is primarily concerned with how to quantitatively deal with imprecision and uncertainty, and offers the decision maker another tool in addition to the classical deterministic and probabilistic mathematical tools that are used in modeling real-world problems. Fuzzy set theory was introduced in the engineering/computer sciences literature by Zadeh (1965) and has gained widespread prominence as a means to model vague data in production management applications. A comprehensive review on the application of fuzzy set theory in production management can be found in Guiffrida and Nagi (1998).

In deterministic inventory models, the analyst must assign a unique numerical value to each model cost parameter as well as to the demand and production rates that characterize the market place and production environment in which the model operates. Similarly, for stochastic inventory models, the assignment process involves using random variables to quantify these inputs under conditions of uncertainty. The advantage of using fuzzy set theory in inventory management lies with its inherent ability to quantify vagueness and imprecision. The purpose of this chapter is to provide a review of the application of fuzzy set theory to inventory management. The review will highlight how fuzzy logic has been used in inventory models and will provide inventory analysts with an alternative, and perhaps complementary, methodology for developing inventory models.

The chapter is organized into two major components. Collectively, Sections 8.2 and 8.3 provide the motivation and foundation for using fuzzy set theory in modeling inventory systems. In Section 8.2, we address the current state of the nonfuzzy modeling efforts for obtaining estimates of the cost parameters of inventory models. The literature on formulating cost estimates for inventory model parameters is reviewed. The robust properties of inventory models to errors in parameter estimation and misspecification of optimal lot sizes are also discussed. In Section 8.3, a review of classical inventory models is presented. In Section 8.4, we build upon this framework and demonstrate a review of the literature on fuzzy inventory models. A summary and conclusion for the chapter is presented in Section 8.5.

8.2   PARAMETER ESTIMATION IN CLASSICAL INVENTORY MODELS

In this section we summarize the literature on determining input values for the inventory cost parameters of order, holding, and shortage. These three parameters are present in nearly all inventory models which have been reported in the literature. The property of robustness is also introduced and material is presented that demonstrates the sensitivity of inventory models to specific errors in the optimal lot size and in the specification of individual parameters.

8.2.1   SUMMARY OF MODELING INVENTORY COST INPUTS

Mathematical inventory models, like all management science/operations research models, require the assignment of numerical values to the input parameters of the model. There is widespread agreement among inventory analysts as to the need of assigning such numerical values; however, the process of estimating the parameters is of concern. Wagner (1980) presents a comprehensive review of the practical problems associated with implementing inventory control systems. A key issue identified in this review is the need for a thoughtful and scientific investigation of the issues relating to the specification of demand forecasts and costs related to inventory modeling.

Deterministic inventory models typically require inputting at least two cost parameters: an order cost for making replenishment and an inventory holding cost for carrying items in inventory. Stochastic inventory models generally require a cost associated with stocking out and back-ordering, in addition to ordering and inventory

TABLE 8.1

Summary of Research on Cost Inputs for Inventory Models

	Model Cost Component		
Research Study	Order	Holding	Stockout
Anderson et al. (2006)			X
Badinelli (1986)			X
Berling (2008)		X	
Biggs et al. (1990)	X		
Bock (1964)		X	
Brown (1985)		X	
Campbell and Rankin (1998)	X		
Chang and Niland (1967)			X
Coopersmith (1990)	X		
Denicoff et al. (1960)			X
Jordan (2008)		X	
Karr (1958)			X
Kim (1991)			X
Lambert (1975)		X	
Lambert et al. (1990)	X		
Malucci (2008)	X		
Oral et al. (1972)			X
Reid et al. (1984)	X		
Schary and Christopher (1979)			X
Schwartz (1966)			X
Walter and Grabner (1975)			X

holding costs. Zanakis et al. (1980) refer to cost estimation in inventory control as the thorn that aggravates the users of inventory control models. Gardner (1980) states that the commonly and almost universally accepted belief held by inventory analysts is that relevant cost information for determining the input values for the parameters of an inventory model can somehow be obtained from a firm's accounting system.

Table 8.1 provides an overview of research studies over the past fifty years that have addressed how to determine the costs associated with ordering, holding, and stockouts in inventory models.

The major concern throughout the studies outlined in Table 8.1 is the capture of accurate and relevant cost data that can be used to describe the costs associated with ordering, holding, and shortages. Today's advancements in inventory control software, such as material requirements and enterprise planning systems, as well as the growing implementation of radio frequency identification, has made the handling and integration of inventory-related information more efficient and available to the decision maker in real-time. However, the troublesome issue of defining numerical values for the input parameters of an inventory model still exists. It must be recognized that totally error-free parameterization of inventory models is not very likely. In the next section we examine the extent to which the robust properties of the inventory models themselves afford protection again parameter misspecification.

8.2.2 ROBUST PROPERTIES OF THE EOQ INVENTORY MODEL

The sensitivity of the inventory cost of the EOQ to errors in the estimation of its model parameters can be found in numerous inventory management textbooks that were published between 1950 and 1975. Starr and Miller (1962, 14) introduced the concept of "sufficient accuracy" with respect to the measurement of inventory-related costs. They note that the typical EOQ-based inventory analysis is not overly sensitive to reasonable errors in the measurements of cost, and that the optimal course of action as suggested by the solution of the inventory model is not likely to be changed very much, even by relatively large errors in the measurements of costs. As such, inventory analysts often take comfort in the knowledge that uncertainty and error introduced when specifying the numerical value of parameters input to an inventory model may have little effect on determining the value of inventory policy variables as a result of the "robust" property inherent to some mathematical inventory models.

Analyses that highlight the robustness of EOQ-based inventory models have developed beyond the scope of inventory management textbooks and into the mainstream inventory management research literature. The research on this topic has developed in two ways. First, researchers have examined the effect of incorrect lot sizes on the total cost performance of the EOQ inventory system; second, researchers have examined the effects of misspecifications of individual and group parameters on the total cost performance of the EOQ inventory system.

8.2.2.1 Misspecification of Optimal Order Quantity

Several researchers have examined the cost penalty associated with the EOQ model when a nonoptimal order quantity (Q) is used in place of the optimal EOQ order quantity (Q^*). Wagner (1969) presents the penalty cost ratio defined by

$$PCP = \frac{1}{2}\left[\frac{Q}{Q^*} + \frac{Q^*}{Q}\right] - 1 \tag{8.1}$$

For example, if the optimal EOQ order quantity is overestimated by 25 percent, the resulting percentage cost error is approximately 2.5 percent. Brown et al. (1986) generalized Equation 8.1 to the form

$$PCP = \left(\frac{1}{n+1}\right)\left[\frac{nQ^*}{Q} + \left(\frac{Q}{Q^*}\right)^n\right] - 1 \tag{8.2}$$

when the relationship between holding costs and average inventory is represented by a power function defined by index n. Diseconomies of scale between holding cost and average inventory are introduced when $n < 1$; economies of scale are introduced when $n > 1$. When $n = 1$, a linear relationship between holding cost and average inventory is implied and Equation 8.2 simplifies to Equation 8.1. Chigopekar (1990) further generalizes this stream of research by introducing planned shortages into the EOQ model with shortage costs modeled by a power function of index n. The resulting penalty cost ratio is

$$PCP = \left(\frac{1}{n+1}\right)\left[\frac{n}{X} + \frac{Y^{n+1}}{X(1-a)} + \frac{(X(1+a)-Y)^{n+1}}{X(1+a)a^n}\right]\qquad(8.3)$$

where X is the ratio of the nonoptimal order quantity (Q) to the optimal order quantity (Q^*), Y is the ratio of the nonoptimal maximum shortage (S) to the optimal maximum shortage (S^*), and $a = (Q/S^*) - 1$. When the deviation in Q^* and S^* are in the same proportion, $X = Y$ and Equation 8.3 simplifies to Equation 8.2. A user-friendly spreadsheet methodology for exploring the robust property of the EOQ to lot size errors using the penalty cost ratio (PCP) approach is found in Guiffrida and Papp (2008).

8.2.2.2 Misspecification of Order Quantity Parameters

Lowe and Schwarz (1983) note that the parameter inputs for the EOQ lot-sizing model are assumed to be stationary and known, when in reality these quantities are often nonstationary and more accurately defined as a function of time or a range of values. They develop two measures for assessing the accuracy (or "goodness") of using parameter estimates in the EOQ model as well as two procedures (minimax and expected value criteria) for estimating the uncertain parameters. For the stationary cases, formulae based on the minimax and expected value criteria are used to develop parameter estimators based on ranges where uncertain parameters are defined by minimum and maximum values. In the nonstationary cases, formulae are developed that introduce time dependency into the range-based parameter estimates. A numerical example is used to demonstrate change in inventory policy resulting from the application of the difference in accuracy measures and estimation procedures.

Dobson (1988) extends the work of Lowe and Schwarz (1983) and demonstrates through a series of propositions a more detailed quantification of the insensitivity of the EOQ inventory model under conditions of parameter uncertainty. Error bounds for the expected error ratio are defined for the joint distribution of the order cost, holding cost, and annual demand parameters of the EOQ. The error bounds are derived for the distribution independent case (e.g., estimates are only dependent on the range endpoints) and for the case when the distribution of the parameters follows the gamma and beta distributions.

Yu (1997) studied the EOQ model under two robust criteria: (1) minimizing the maximum total inventory cost, and (2) minimizing the maximum percentage deviation from the optimal total inventory cost. Under these criteria, the input data of the EOQ model are defined by continuous intervals. A linear time algorithm was developed to find robust decisions under each criterion. A stochastic EOQ model, in which the model parameter inputs are defined by known probability distributions, is formulated and solved, and the results of the stochastic optimization are compared to the solutions achieved under the robust decision criteria. Theorems are presented that show that the robust decision criteria outperform the stochastic optimization approach with respect to the EOQ inventory model.

8.2.3 LIMITATIONS OF ROBUSTNESS IN THE EOQ MODEL

The scope of robustness with respect to the parameterization of inventory input parameters may be limited for several reasons. First, the properties of robustness have only been studied in the literature for the EOQ inventory model. Second, inventory experts have expressed disagreement as to the assurances that robustness offers in the parameterization of inventory models, such as the EOQ. Peterson and Silver (1979) state that the robustness of the EOQ model is an important fundamental concept in operational inventory control. Zangwill (1987) takes the position that reliance on the robust properties of the EOQ model can lead to false or misleading results, especially when viewed from the zero inventory philosophy.

8.3 THE INVENTORY MANAGEMENT LITERATURE

A comprehensive review of all inventory models published since 1913 has never been undertaken and would be a daunting task due to the sheer number of published inventory models. For example, a search using the phrase "mathematical inventory models" in Google Scholar returns over 90,000 hits. In this section we present a summary of literature reviews on inventory models that have appeared in the literature. In Section 8.3.1 a review of general inventory models is presented. This section covers models that were published before 1982. Section 8.3.2 contains a review of inventory models with special characteristics. This section covers models published since 1982.

8.3.1 OVERVIEW OF GENERAL INVENTORY MODELS

Bramson (1962) presents a review of inventory models with variable lead time. Models are classified according to two coding schemes. The first coding scheme consists of three variables that describe the type of inventory control policies used in the inventory model to govern order review, size, and treatment of shortages. Within this first coding scheme, the order review policies encoded are: (1) continuous review, (2) periodic review, (3) random review, and (4) other/irrelevant. Order sizes are coded as: (1) $Q = 1$, (2) optimal Q, (3) a fixed multiple of Q, and (4) other/irrelevant. Demand under a shortage is treated as: (1) completely back-ordered, (2) completely lost, (3) partially back-ordered and partially lost, and (4) other/irrelevant.

The second coding scheme consists of one to three letters and is used to classify the type of distribution used for demand and lead time. A single letter signifies the form of the distribution used in the model to define customer demand. A second letter identifies the accompanying lead-time distribution. Demand and lead-time distributions considered are: (1) normal, (2) Poisson, (3) exponential/geometric, (4) Erlang, (5) gamma, (6) negative binomial, (7) unit demand, or (8) other. When all three letters are used, the third letter denotes the time interval between demands.

The coding schemes are used to classify ten inventory models published in the literature from 1955 to 1961. Pursuant to the theme of this review paper, Bramson (1962) provides a step-by-step derivation of an expression for comparing the variability of the demand and lead-time distributions for variable demand inventory models in an appendix to the paper.

Aggarwal (1974) provides a comprehensive classification of the literature in mathematical inventory control models that appeared from 1964 to 1973. The classification scheme used to organize inventory models in the paper is identified using a well-designed and easy to read figure found on page 444 of the paper. Within this classification scheme, inventory systems are broadly classified as either static or dynamic models. The static model subgroup is further broken down as to whether the models represent single or multiple items. Both model forms (single or multiple items) are further partitioned based on the number of locations and echelons the models engage. Models in the dynamic subgroup are broken down as to whether the model is deterministic or stochastic. Stochastic models are further classified as to whether they possess a known or unknown demand distribution.

A list of eleven quantifiable inventory variables present in the literature is then discussed, to which Aggarwal (1974) adds four additional variables that cannot be easily quantified. Individual research papers from the inventory literature are grouped into six specific categories. These categories are: (1) models for determining optimum inventory policies (twenty-three papers), (2) lot-size optimization (five papers), (3) optimization of specific management objectives (eight papers), (4) models for optimizing highly specific situations (twenty papers), (5) applications of advanced mathematical theories (nine papers), and (6) models bridging the gap between theory and practice (five papers). The author concludes the review paper with an interesting discussion of the limitations set forth for the general practicability of inventory models.

Two additional inventory review papers that present detailed classification schemes for inventory models are found in Nahmias (1978) and Silver (1981). The classification schemes presented in these papers differ from those of Bramson (1962) and Aggarwal (1974), in that models for perishable products are now included as a part of the overall inventory model classification scheme.

8.3.2 OVERVIEW OF INVENTORY MODELS WITH SPECIAL CHARACTERISTICS

Nahmias (1982) reviewed the literature on inventory models for perishable inventories. The review is restricted to the ordering policies of perishables. Ordering policies for inventory that is subject to obsolescence is excluded. Prior to being introduced as a distinct class of inventory model in the classifications schemes presented by Nahmias (1978) and Silver (1981), selected perishable inventory models published prior to 1973 were reviewed by Aggarwal (1974) under the classification of "models for optimizing highly specific situations."

The classification scheme used by Nahmias (1982) is based on two major groupings: (1) models for single and multiple products with fixed lifetime perishability under conditions of deterministic and stochastic demand, and (2) models with random lifetime perishability in which the product lifetime follows either a known probability distribution or experiences exponential decay. Similar in form to his prior review paper (Nahmias 1978), a detailed and thorough explanation of each class of model addressed in the review is provided. Thirty papers from the literature are reviewed under the category of models with fixed life perishability; nine papers are reviewed under the category of random lifetime perishability. Noting the analogy between queuing models with customer impatience and perishable inventory models, two papers that apply analytical results from the queuing application are examined. The review concludes with the acknowledgment that interest in perishable inventory problems has been influenced by problems in blood bank management, and selected papers from the literature on blood bank management are reviewed.

Raafat (1991) presents a review of the literature on continuously deteriorating mathematical inventory models (also known as inventory models subject to exponential decay or age-independent perishable inventory models). Following the classification scheme defined in Silver (1981), eight groupings are defined for categorizing the literature on continuously deteriorating inventory models. Within seven of the groupings, two distinct subgroupings are introduced to further partition models found in the literature. Nine classes of continuously deteriorating inventory models are identified and each class of model is then categorized according to the classification scheme defined in the paper. Within a given model class, relevant papers from the literature are addressed. A total of sixty-three papers are reviewed across the nine model classes.

Benton and Park (1996) reviewed the literature on inventory lot sizing models with quantity discounts. The taxonomy adopted consists of three nested classifications. The first classification categorizes quantity discount models according to whether the demand profile underlying the model is time phased. Second, within each demand class the type of discount model (all-units discount or incremental discount) is established. Last, the model is classified from the point of view of the parties involved (e.g., from the lot sizing perspective of the buyer only or from the joint lot sizing perspective of both the buyer and supplier).

A total of fifty-one papers on inventory lot sizing models with quantity discounts were reviewed. The vast majority of the papers incorporated models that were not time phased with respect to demand (thirty-five papers). Forty papers had models that were based on incremental discount schedules; eleven papers had models that were based on all-units discount schedules. The vast majority of the papers did not adopt a joint buyer–supplier orientation with respect to lot sizing (thirty-six papers), and no papers addressed the buyer–supplier interaction under a time-phased demand orientation. Benton and Park (1996) acknowledge that more complex models are needed to address the buyer–seller interaction with respect to the lot sizing decision involving quantity discounts when demand is time phased.

Goyal and Giri (2001) present a review of the literature on deteriorating inventory models published since the early 1990s. Their review addresses new contributions to the literature on deteriorating inventory models that have been published since the earlier review of the field Raafat (1991). Their review is classified according to three categories: (1) models for inventory with fixed lifetimes, (2) models for inventory with a random life, and (3) models for inventory with decays corresponding to the proportional inventory decrease in terms of its utility or physical quantity. The detail of the review effort is impressive with consideration given to the demand structures used

TABLE 8.2
Summary of Literature Reviews on Specialized Topics in Inventory Management

Research Study	Inventory Topic Reviewed
Khouja (1999)	Newsvendor models
Kennedy et al. (2002)	Spare-parts inventory
Minner (2003)	Inventory models with multiple suppliers
Urban (2005)	Inventory-level-dependent demand models
Silver (2008)	Inventory models contributed by Canadian researchers
Williams and Tokar (2008)	Inventory models published in logistics journals
Ben-Daya et al. (2008)	Joint buyer–vendor inventory models

by researchers in modeling deteriorating inventory, as well the specialized model attributes that address pricing, scheduling of payments, the time value of money, and inventory storage. The review encompasses more than 120 new contributions to the literature on deteriorating inventory models.

Several authors have published recent reviews of specialized topics within the inventory management literature. These reviews are summarized in Table 8.2.

8.4   FUZZY INVENTORY LITERATURE

The literature survey on mathematical inventory models that was presented in Section 8.3 provides a foundation for examining the literature on fuzzy inventory models. In this section we will review the literature on fuzzy inventory models. The review is organized along seven general classes of inventory models and their extensions. These classes are economic order quantity models, economic production quantity models, joint economic lot sizing models, single period newsvendor models, multiperiod inventory models, multiproduct inventory models, and systems (continuous and periodic review) for inventory control.

The objective of this survey is to identify how fuzzy set theory has been used in the formulation of the fuzzy inventory model. The semantics of fuzzy set theory will not be discussed as a part of this review. The interested reader is referred to Kaufmann and Gupta (1988) for the supporting details on the mathematics of fuzzy inference. An introduction on how to use fuzzy set theory in mathematical inventory models is found in Buckley et al. (2002) and Buckley (2005). These references expertly illustrate how to apply fuzzy set theory in the formulation and solution of inventory models.

8.4.1   Economic Order Quantity (EOQ) Models

As in the traditional inventory literature, the fuzzy economic order quantity has provided the genesis for the evolution of fuzzy inventory models. Six papers have addressed the basic EOQ model under varying schemes of fuzzy parameter

TABLE 8.3
Fuzzy Economic Order Quantity Models

Economic Order Quantity Model

Model	Order Quantity	Model Input Parameters		
		Order Cost	Holding Cost	Annual Demand
Park (1987)	C	F	F	C
Vujosevic (1996)	C	F	F	C
Lee and Yao (1999)	F	C	C	C
Yao et al. (2000)	F	C	C	F
Yao and Chiang (2003)	C	C	F	F
Wang et al. (2007)	C	F	F	C

Economic Order Quantity Model with Back Orders

Model	(Q, S)	Model Input Parameters			
		Order Cost	Holding Cost	Back-Order Cost	Annual Demand
Chen et al. (1996) Yao and Lee (1996)	(C, C)	F	F	F	F
Yao and Lee (1999)	(F, C)	C	C	C	C
Chang et al. (1998)	(C, F)	C	C	C	C
Yao and Su (2000)	(C, C)	C	C	C	F
Wu and Yao (2003)	(F, F)	C	C	C	C
Yao and Su (2008)	F max inv	C	C	C	F

Note: C = crisp (nonfuzzy) model parameter/attribute. F = fuzzy model parameter/attribute.

assignments. An additional seven papers model the fuzzy EOQ with back orders (see Table 8.3). Ten papers provide extensions to the fuzzy EOQ model (see Table 8.4).

We introduce the following notation for use throughout this section. Let C denote a crisp (nonfuzzy) model parameter/attribute; let F denote a fuzzy model parameter/attribute.

8.4.2 ECONOMIC PRODUCTION QUANTITY (EPQ) MODELS

The economic order quantity model assumes that all units of the EOQ lot size are delivered from an outside vendor at the same time. Hence, if the EOQ was used as an internal production lot sizing model, the assumption of instantaneous replenishment would equate to an infinite (and unrealistic) production rate. By relaxing the instantaneous replenishment to allow for a production rate that is greater than the demand rate, the EOQ model can be modified to determine the optimal production lot size as a function of setup and inventory holding costs. This modification of the traditional

TABLE 8.4

Extensions to the Fuzzy Economic Order Quantity Model

Model	Extension of Traditional EOQ Model	Fuzzy Model Attributes
Roy and Maiti (1995)	Extends the classic EOQ model to include a constraint for available warehouse storage space.	Ordering costs, holding costs, and the maximum available warehouse storage space are defined using fuzzy numbers.
Mondal and Maiti (2002)	Formulates a multiple item EOQ model under constraints for total investment, storage space, and the number of production runs.	Two fuzzy model formulations are presented: In model 1, the available storage area and maximum number of orders are defined by fuzzy numbers. In model 2, ordering costs and inventory holding costs are defined by fuzzy numbers.
Chang (2003)	Extends Porteus' (1986) EOQ model for process quality and setup reduction quality improvement.	The opportunity cost of capital for process improvement is modeled as a fuzzy number.
Chang (2004)	Extends the order quantity model of Salameh and Jaber (2000) for items of imperfect quality.	Order quantity models for items of imperfect quality are presented when the defect rate and annual demand are defined by fuzzy numbers.
Yadavalli et al. (2005)	Formulates a multiple item EOQ model subject to a budget constraint for the number of stocked units.	Order costs and inventory holding costs are defined by fuzzy numbers.
De and Goswami (2006)	Develops an EOQ model under conditions of inflation and product deterioration for a finite time horizon with permissible payment delay.	The deterioration rate and inflation rate are modeled as fuzzy numbers.
Chen and Ouyang (2006)	Extends the order quantity model of Jamal et al. (1997) for deteriorating items with allowable shortage and permissible delay in payment.	The inventory carrying cost and the interest rate (earned and paid) are defined as fuzzy numbers.
Mahata and Goswami (2007)	Extends the EOQ model to include permissible delays in payments to both retailers and customers under conditions of product deterioration.	The demand rate and inventory costs of setup, holding, and purchasing are defined as fuzzy numbers.
Liu (2008)	Redefines the EOQ as a profit maximization model where demand is a function of the unit item price, and the item unit cost is a decreasing function of the order quantity.	Annual demand and the item unit cost are defined as fuzzy numbers.
Roy et al. (2008)	Formulates a multiple product EOQ model for deteriorating items subject to constraints on cost and storage space.	Ordering cost, inventory holding cost, available storage space, and budgetary cost are defined by fuzzy numbers.

TABLE 8.5
Fuzzy Economic Production Quantity Model

Model	Production Lot Size	Demand	Production	Setup Cost	Inventory Holding Cost
				Model Input Parameters	
Lee and Yao (1998)	C	F	F	C	C
Chang (1999)	F	C	C	C	C
Lin and Yao (2000)	F	C	C	C	C
Hsieh (2002)					
Model 1	C	F	F	F	F
Model 2	F	F	F	F	F

Note: C = crisp (nonfuzzy) model parameter/attribute. F = fuzzy model parameter/attribute.

EOQ model is often referred to as the economic production quantity (EPQ) model. An early treatment of the formulation of the EPQ model is found in Eilon (1956).

Tables 8.5 and 8.6 overview twelve fuzzy EPQ models have been identified in the literature. Four papers are based on the traditional (crisp) EPQ model and use fuzzy numbers to model the input parameters and the optimal production lot size; eight of the papers apply fuzzy logic to extensions of the traditional EPQ model that addresses multiple items, constraints resulting from cost budgets and limited storage space, and imperfect product quality.

8.4.3 JOINT ECONOMIC LOT SIZING MODELS

Inventory models that address issues of inventory coordination between a buyer and seller have been extensively studied in the literature. This class of inventory models is commonly referred to as joint economic lot sizing (JELS) models. The objective of these models is the development of a jointly coordinated buyer–seller inventory strategy that is more beneficial to each member's individual noncoordinated inventory strategy. A detailed review of the literature on crisp model formulations of the joint economic lot size model is found in Ben-Daya et al. (2008).

Table 8.7 summarizes six fuzzy JELS models that have been introduced, with each fuzzy model representing an extension to a previously published JELS model.

8.4.4 SINGLE PERIOD, NEWSVENDOR MODELS

The newsvendor model is a single-period, probabilistic inventory model whose objective is to determine the order quantity that minimizes expected underage costs (costs due to shortage) and overage costs (costs due to holding inventory). Alternatively, the model can also be formulated to maximize expected profits. The interest in the newsvendor model is high with more than fifty papers published on applications of

TABLE 8.6
Extensions to the Fuzzy Economic Production Quantity Model

Model	Extension of Traditional EPQ Model	Fuzzy Model Attributes
Pappis and Karacapilidis (1995)	Extends the EPQ model to include multiple items using the common cycle method of Buffa and Sarin (1987).	The optimal number of production runs for a set of products produced in a batch production environment is determined when the aggregate demand across all products is modeled as a fuzzy number.
Roy and Maiti (1997)	A demand-dependent unit cost and quantity-dependent setup cost are introduced subject to a constraint on storage capacity. The model is formulated for both single and multiple items.	The optimal production lot size is modeled and solved as a fuzzy mathematical programming model that minimizes the total cost of setup, holding, and production subject to a fuzzy storage capacity constraint.
Chang and Chang (2006)	Includes a production cost term and incorporates a unit cost structure that is lot-size dependent.	The production lot size, demand rate, production rate, and item unit cost are defined using fuzzy numbers.
Mandal and Roy (2006a)	Develops a multiple-item production lot sizing model were items of imperfect quality are subject to repair.	Production repair, setup, and holding costs are modeled as fuzzy numbers.
Chang et al. (2006)	Extends the EPQ to include multiple items.	Item demands are defined as fuzzy numbers.
Chen et al. (2007)	Allows imperfect products that can be sold at a discount.	The production lot size and cost inputs of the model are defined using fuzzy numbers.
Islam and Roy (2007)	Formulates a multiple-item EPQ model with demand-dependent unit production costs and production process reliability subject to an inventory space constraint.	Production, inventory carrying, interest, and depreciation costs as well as the available storage space are modeled as fuzzy numbers.
Panda et al. (2008)	Incorporates fuzzy budget and shortage constraints to the multiproduct EPQ model with imperfect production.	The maximum value of the budgets for total production and screening costs, and for total shortage costs are defined as fuzzy numbers.

the model in inventory control. A detailed review of the literature up to 1999 was conducted by Khouja (1999).

Fuzzy extensions to the newsvendor model are found in nine papers (see Table 8.8). Early papers (Petrovic et al., 1996; Ishii and Konno, 1998; Li et al., 2002) concentrated on using fuzzy numbers as an alternative to modeling demand as a random variable and for defining potentially imprecise cost-input parameters. Later papers (Kao and Hsu, 2002a; Dutta et al., 2005; Shao and Ji, 2006) concentrated on demonstrating how specialized solution procedures from fuzzy set theory could be used

TABLE 8.7

Fuzzy Joint Economic Lot Sizing Models

Fuzzy JELS Model	JELS Model Extended	Fuzzy Model Attributes
Lam and Wong (1996)	Dolan (1978)	Single and multiple incremental price discounts are modeled as fuzzy numbers.
Das et al. (2004a)	Yang and Wee (2000)	A fuzzy multiobjective model is developed for a deteriorating item that is subject to a price discount. A fuzzy goal programming methodology is used to solve the model.
Mahata et al. (2005)	Banerjee (1986)	The joint buyer–vendor lot size is modeled as a fuzzy number.
Ouyang et al. (2006)	Goyal and Nebebe (2002)	Defective items are introduced into the JELS formulation and modeled as a fuzzy number.
Yang (2007)	Yang (2006)	The annual demand and adjustable production rate are modeled as fuzzy numbers.
Sinha and Sarmah (2008)	Sinha and Sarmah (2007)	The annual demand as well as the order and holding costs of the buyer are modeled as fuzzy numbers.

to solve the fuzzy newsvendor problem. Most current papers (Yao et al., 2006; Dutta et al., 2007a; Dey and Chakraborty, 2008) investigate general extensions of the fuzzy newsvendor model into areas of cash management, multiple procurement opportunities, and resalable returns.

8.4.5 MULTIPERIOD INVENTORY MODELS

In real-world applications, inventory and production decisions are interdependent and temporal in nature. Fuzzy logic has been useful in formulating multiperiod lot sizing models. Table 8.9 provides an overview of fuzzy multiperiod inventory models.

8.4.6 MULTI-ITEM INVENTORY MODELS

Multi-item models make up a large number of fuzzy inventory models found in the literature. These models typically require advanced solution methodologies such as genetic algorithms, fuzzy simulation-based optimization, and fuzzy nonlinear programming. A summary of fuzzy multi-item inventory models is found in Table 8.10.

8.4.7 INVENTORY CONTROL SYSTEMS

The trade-offs associated with using continuous review versus periodic review inventory control systems are well addressed in the inventory literature (see, for example, Silver et al., 1998). The majority of inventory models reported in the fuzzy inventory literature have adopted a continuous review inventory control system (see Table 8.11).

TABLE 8.8
Fuzzy Newsvendor Models

Model	Demand	Costs Purchase	Holding	Shortage	Key Model Attributes
Petrovic et al. (1996)					First to formulate a fuzzy newsvendor model. Presents two cost-minimization
Model 1	F	C	C	C	models that extend the classic discrete
Model 2	C	C	F	F	demand newsvendor inventory model.
Ishii and Konno (1998)	C	C	C	F	Profit maximization of the classic discrete demand newsvendor model.
Li et al. (2002)					Demand is modeled as a continuous
Model 1	C	C	F	F	fuzzy random variable in model 2.
Model 2	F	C	C	C	
Kao and Hsu (2002a)	F	C	C	C	Utilizes an area measurement solution index based on the fuzzy number ranking procedure of Yager (1981) to solve the fuzzy newsvendor model.
Dutta et al. (2005)	F	C	C	C	Demand is modeled as a continuous fuzzy random variable and the fuzzy newsvendor model is solved using the graded mean integration method of Chen and Hsiesh (1999).
Shao and Ji (2006)	F	C	C	C	Extends the newsvendor model with fuzzy demand to the multiproduct case under a profit maximization objective. A hybrid genetic algorithm/fuzzy simulation solution methodology is used to solve the model.
Yao et al. (2006)	F	C	C	C	Extends the cash management newsvendor model of Johnson and Montgomery (1974) to include fuzzy demand.
Dutta et al. (2007a)	F	C	C	C	Extends the fuzzy newsvendor model to accommodate two procurement opportunities within the single period horizon.
Dey and Chakraborty (2008)	F	C	C	C	Extends the resalable returns newsvendor model of Vlachos and Dekker (2003) to include fuzzy demand.

Note: C = crisp (nonfuzzy) model parameter/attribute. F = fuzzy model parameter/attribute.

TABLE 8.9
Fuzzy Multiperiod Inventory Models

Model	Key Model Attributes
Sommer (1981)	Uses fuzzy dynamic programming to determine optimal inventory and production levels in a real-world integrated multiperiod inventory and production scheduling problem for an organization engaged in a planned withdrawal from a market.
Kacprzyk and Staniewski (1982)	Applies fuzzy set theory to determine an optimal aggregate inventory replenishment strategy subject to a set of long-term management objectives.
Lee et al. (1990)	Introduces fuzzy logic into material requirements planning (MRP) by defining period demand as a fuzzy number. A fuzzy part period balancing algorithm is developed.
Lee et al. (1991)	Extends their previous research on multiperiod fuzzy lot sizing and introduces fuzzy versions of the Wagner–Whitin and Silver–Meal lot sizing models.
Liu (1999)	Applies fuzzy decision making to investigate optimal inventory policy for a multiperiod inventory system with partial back orders.

TABLE 8.10
Fuzzy Multi-Item Inventory Models

Model	Key Model Attributes
Roy and Maiti (1998)	Formulates a multiobjective inventory model for multiple items subject to deterioration and with stock-dependent demand. The objectives of the model are maximizing profit and minimizing wastage cost. Fuzzy numbers are used to define the profit goal, wastage cost, and storage area.
Das et al. (2000)	Develops a multi-item inventory model with demand-dependent unit cost subject to fuzzy constraints on total storage area, shortage cost, and average inventory investment.
Yao et al. (2003)	Models the optimal ordering policy for two mutually complementary products when the item prices and order quantities are fuzzy.
Das et al. (2004b)	Formulates a multi-item stochastic inventory model with back orders and with demand-dependent unit costs subject to fuzzy budgetary and storage space constraints.
Mandal et al. (2005)	Develops a multi-item, multiobjective inventory model with shortages and demand-dependent unit costs subject to fuzzy constraints on storage space, number of orders, and production cost. Inventory holding, order, and shortage costs are defined by fuzzy numbers.
Maiti and Maiti (2006)	Formulates a multi-item inventory model where item demand is dependent upon the selling price, frequency of advertisement, and displayed inventory level of the item. The objective of the model is to maximize profit subject to storage in two storehouses and investment constraints. Fuzzy numbers are used to define the storehouse capacity, the item purchase cost, and maximum inventory investment.

(Continued)

TABLE 8.10 (CONTINUED)
Fuzzy Multi-Item Inventory Models

Model	Key Model Attributes
Mandal and Roy (2006b)	Introduces a multi-item inventory model for items that are displayed on two shelves (back room and front room) and item demand is a function of the displayed inventory level. The model objective determines the optimal display and order quantity of each item such that average profit is maximized. The available display area and item purchase, holding, shelf, and order costs are defined by fuzzy numbers.
Maiti and Maiti (2007b)	Formulates a multi-item inventory model for seasonal products where item demand is a linearly increasing function of time. Upon sale, items are governed by a bulk release pattern between the two warehouses. Item ordering costs and shortage costs are defined by fuzzy numbers.
Roy et al. (2007)	Develops a multi-item model to determine the optimal item production and inventory levels subject to fuzzy constraints on the maximum available storage space and the total budgetary capital.
Maiti (2008)	Presents a multi-item inventory model that includes inflation, the time value of money, and a bulk storage pattern over two warehouses. Item purchase costs, the maximum inventory investment, and the maximum storage at the source warehouse are defined by fuzzy numbers.

TABLE 8.11
Fuzzy Inventory Control Systems

Model	Type of Control System	Fuzzy Model Attributes
Gen et al. (1997)	Continuous review	Three models under fuzzy ordering, inventory holding, and shortage costs are considered: (1) crisp order quantity and reorder point, (2) fuzzy order quantity and crisp reorder point, and (3) crisp order quantity and fuzzy reorder point.
Katagiri and Ishii (2000) Katagiri and Ishii (2002)	Periodic review	Extends the optimal rotation and allocation model for perishable products of Nose et al. (1981) to include fuzzy shortage and outdating costs.
Petrovic and Petrovic (2001)	Continuous review, periodic review	Examines three inventory performance measures (fill rate, inventory holding cost, and regularity of replenishment) under both continuous and periodic review polices when annual demand is defined by a fuzzy number.
Samanta and Al-Araimi (2001)	Periodic review	Presents a fuzzy logic–based model for inventory control. Simulation is used to illustrate the inventory control capabilities of the model.
Ouyang and Chang (2001)	Continuous review	Extends the variable lead time and partial back-order model of Moon and Choi (1998) to have a fuzzy lost sales rate.

(Continued)

TABLE 8.11 (CONTINUED)
Fuzzy Inventory Control Systems

Model	Type of Control System	Fuzzy Model Attributes
Kao and Hsu (2002b)	Continuous review	Determines the optimal order quantity and reorder point when lead time demand is modeled with crisp lead time and fuzzy demand.
Ouyang and Yao (2002)	Continuous review	Implements a minimax distribution free solution procedure to determine the optimal order quantity and reorder point when lead time demand is modeled with variable lead time and fuzzy demand.
Ouyang and Chang (2002)	Continuous review	Extends Ouyang and Chang (2001) by incorporating a minimax distribution-free-based solution procedure to determine the optimal fuzzy inventory strategy.
Pai and Hsu (2003)	Continuous review	Determines the optimal order quantity and reorder point when lead time demand and inventory holding cost are fuzzy numbers.
Chang et al. (2004)	Continuous review	Extends the variable lead time model with back orders and lost sales of Ouyang et al. (1996) to include fuzzy lead time demand and a fuzzy back-order rate.
Dey et al. (2005)	Periodic review	Determines the optimal inventory policy for a multiobjective inventory model when lead time demand, and ordering, holding, and shortage costs are defined as fuzzy numbers.
Chang et al. (2006)	Continuous review	Extends Ouyang et al. (1996) to determine the optimal order quantity and lead time for fuzzy lead time demand and fuzzy annual demand.
Maiti and Maiti (2007a)	Continuous review	Formulates and solves, using a genetic algorithm, the optimal order quantity and reorder point variables for a warehouse inventory model with back orders when lead time demand is fuzzy.
Dutta et al. (2007b)	Continuous review	Determines the optimal order quantity and reorder point when lead time demand and annual demand are defined by fuzzy numbers.
Lin (2008)	Periodic review	Extends the periodic review model with variable lead time of Ouyang and Chuang (2001) to include a fuzzy shortage quantity and fuzzy back-order rate.
Vijayan and Kumaran (2008)	Continuous review, periodic review	Optimal inventory policies are determined for models when the costs associated with ordering, holding, and shortage are jointly and individually defined by fuzzy numbers.
Tutuncu et al. (2008)	Continuous review	Introduces a decision support system to determine the optimal order quantity and reorder point when the costs of production, ordering, inventory holding, and storage are defined by fuzzy numbers.

8.5   SUMMARY AND CONCLUSIONS

This chapter has addressed an extensive literature review and survey of fuzzy set theory in inventory management. Eighty-three inventory models organized into seven general classes of inventory models have been reviewed. The emphasis in each review was to identify how fuzzy set theory was used in the formulation of the inventory model.

The chapter has also provided a summary of published literature reviews on traditional inventory models. Issues concerning the parameterization of traditional inventory models were also discussed. It is against this landscape that researchers in inventory management may identify the potential benefits of using fuzzy set theory. This chapter should give researchers in inventory management new tools and ideas on how to approach modeling problems in inventory management. The chapter also provides a foundation for researchers currently engaged in research of fuzzy inventory models, and hopefully may identify and stimulate new advancements in the application of fuzzy set theory to problems in inventory management.

REFERENCES

Fuzzy Literature

Buckley, J. J. 2005. *Fuzzy probabilities: New approach and applications*. Berlin/Heidelberg: Springer.

Buckley, J. J., T. Feuring, and Y. Hayashi. 2002. Solving fuzzy problems in operations research: Inventory control. *Soft Comput* 7(2):121–129.

Chang, H.-C. 2003. Fuzzy opportunity cost for EOQ model with quality improvement investment. *Int J Syst Sci* 34(6):395–402.

Chang, H.-C. 2004. An application of fuzzy sets theory to the EOQ model with imperfect quality items. *Comput Oper Res* 31(12):2079–2092.

Chang, H.-C., J.-S. Yao, and L.-Y. Ouyang. 2004. Fuzzy mixture inventory model with variable lead-time based on probabilistic fuzzy set and triangular fuzzy number. *Math Comput Model* 39(2/3):287–304.

Chang, H.-C., J.-S. Yao, and L.-Y. Ouyang. 2006. Fuzzy mixture inventory model involving fuzzy random variable lead time demand and fuzzy total demand. *Eur J Oper Res* 169(1):65–80.

Chang, P.-T., and C.-H. Chang. 2006. An elaborative unit cost structure-based fuzzy economic production quantity model. *Math Comput Model* 43(11/12):1337–1356.

Chang, P.-T., M.-J. Yao, S.-F. Huang, and C.-T. Chen. 2006. A genetic algorithm for solving a fuzzy economic lot-size scheduling problem. *Int J Prod Econ* 102(2):265–288.

Chang, S.-C. 1999. Fuzzy production inventory for fuzzy product quantity with triangular fuzzy number. *Fuzzy Set Syst* 107(1):37–57.

Chang, S.-C., J.-S. Yao, and H.-M. Lee. 1998. Economic reorder point for fuzzy backorder quantity. *Eur J Oper Res* 109(1):183–202.

Chen, L.-H., and L.-Y. Ouyang. 2006. Fuzzy inventory model for deteriorating items with permissible delay in payment. *Appl Math Comput* 182(1):711–726.

Chen, S. H., and C. H. Hsieh. 1999. Graded mean integration representation of generalized fuzzy number. *J Chin Fuzzy Syst Assoc* 5:1–7.

Chen, S. H., C.-C. Wang, and S. M. Chang. 2007. Fuzzy economic production quantity model for items with imperfect quality. *Int J Innovat Comput Inform Contr* 3(1):85–95.

Chen, S.-H., C.-C. Wang, and A. Ramer. 1996. Backorder fuzzy inventory model under function principle. *Inform Sci* 95(1/2):71–79.

Das, K., T. K. Roy, and M. Maiti. 2000. Multi-item inventory model with quantity-dependent inventory costs and demand-dependent unit cost under imprecise objective and restrictions: A geometric programming approach. *Prod Plann Contr* 11(8): 781–788.

Das, K., T. K. Roy, and M. Maiti. 2004a. Buyer-seller fuzzy inventory model for a deteriorating item with discount. *Int J Syst Sci* 35(8):457–466.

Das, K., T. K. Roy, and M. Maiti. 2004b. Multi-item stochastic and fuzzy-stochastic inventory models under two restrictions. *Comput Oper Res* 31(11):1793–1806.

De, S. K., and A. Goswami. 2006. An EOQ model with fuzzy inflation rate and fuzzy deterioration rate when a delay in payment is permissible. *Int J Syst Sci* 37(5):323–335.

Dey, J.-K., S. Kar, and M. Maiti. 2005. An interactive method for inventory control with fuzzy lead-time and dynamic demand. *Eur J Oper Res* 167(2):381–397.

Dey, O., and D. Chakraborty. 2008. A single-period inventory problem with resalable returns: A fuzzy stochastic approach. *Int J Math Phys Eng Sci* 1(1):8–15.

Dutta, P., D. Chakraborty, and A. R. Roy. 2005. A single-period inventory model with fuzzy random variable demand. *Math Comput Model* 41(8/9):915–922.

Dutta, P., D. Chakraborty, and A. R. Roy. 2007a. An inventory model for single-period products with reordering opportunities under fuzzy demand. *Comput Math Appl* 53(10):1502–1517.

Dutta, P., D. Chakraborty, and A. R. Roy. 2007b. Continuous review inventory model in mixed fuzzy and stochastic environment. *Appl Math Comput* 188(1):970–980.

Gen, M., Y. Tsujimura, and D. Zheng. 1997. An application of fuzzy set theory to inventory control models. *Comput Ind Eng* 33(3/4):553–556.

Guiffrida, A. L., and R. Nagi. 1998. Fuzzy set theory applications in production management research: A literature review. *J Intell Manuf* 9(1):39–56.

Hsieh, C. H. 2002. Optimization of fuzzy production inventory models. *Inform Sci* 146(1/4):29–40.

Ishii, H., and T. Konno. 1998. A stochastic inventory problem with fuzzy shortage cost. *Eur J Oper Res* 106(1):90–94.

Islam, S., and T. K. Roy. 2007. Fuzzy multi-item economic production quantity model under a space constraint: A geometric programming approach. *Appl Math Comput* 184(2): 326–335.

Kacprzyk, J., and P. Staniewski. 1982. Long term inventory policy-making through fuzzy decision making. *Fuzzy Set Syst* 8:117–132.

Kao, C., and W.-K. Hsu. 2002a. A single period inventory model with fuzzy demand. *Comput Math Appl* 43(6/7):841–848.

Kao, C., and W.-K. Hsu. 2002b. Lot size-reorder point inventory model with fuzzy demands. *Comput Math Appl* 43(10/11):1291–1302.

Katagiri, H., and H. Ishii. 2000. Some inventory problems with fuzzy shortage cost. *Fuzzy Set Syst* 111(1):87–97.

Katagiri, H., and H. Ishii. 2002. Fuzzy inventory problems for perishable commodities. *Eur J Oper Res* 138(3):545–553.

Kaufmann, A., and M. M. Gupta. 1988. *Fuzzy mathematical models in engineering and management science.* Amsterdam: North-Holland.

Lam, S. M., and D. S. Wong. 1996. A fuzzy mathematical model for joint economic lot size problem with multiple price breaks. *Eur J Oper Res* 95(3):611–622.

Lee, H.-M., and J.-S. Yao. 1998. Economic production quantity for fuzzy demand quantity and fuzzy production quantity. *Eur J Oper Res* 109(1):203–211.

Lee, H.-M., and J.-S. Yao. 1999. Economic order quantity in fuzzy sense for inventory without backorder model. *Fuzzy Set Syst* 105(1):13–31.

Lee, Y. Y., B. A. Kramer, and C. L. Hwang. 1990. Part-period balancing with uncertainty: A fuzzy sets theory approach. *Int J Prod Res* 28(10):1771–1778.

Lee, Y. Y., B. A. Kramer, and C. L. Hwang. 1991. A comparative study of three lot-sizing methods for the case of fuzzy demand. *Int J Oper Prod Manag* 11(7):72–80.

Li, L., S. N. Kabadi, and K. P. K. Nair. 2002. Fuzzy models for single-period inventory problem. *Fuzzy Set Syst* 132(3):273–289.

Lin, D.-C., and J.-S. Yao. 2000. Fuzzy economic production for production inventory. *Fuzzy Set Syst* 111(3):465–495.

Lin, Y.-J. 2008. A periodic review inventory model involving fuzzy expected demand short and fuzzy backorder rate. *Comput Ind Eng* 54(3):666–676.

Liu, B. 1999. Fuzzy criterion models for inventory systems with partial backorders. *Ann Oper Res* 87:117–126.

Liu, S.-T. 2008. Fuzzy profit measures for a fuzzy economic order quantity model. *Appl Math Model* 32(10):2076–2086.

Mahata, G. C., and A. Goswami. 2007. An EOQ model for deteriorating items under trade credit financing in the fuzzy sense. *Prod Plann Contr* 18(8):681–692.

Mahata, G. C., A. Goswami, and D. K. Gupta. 2005. A joint economic-lot-size model for purchaser and vendor in fuzzy sense. *Comput Math Appl* 50(10/12):1767–1790.

Maiti, M. K. 2008. Fuzzy inventory model with two warehouses under possibility measure on fuzzy goal. *Eur J Oper Res* 188(3):746–774.

Maiti, M. K., and M. Maiti. 2006. Fuzzy inventory model with two warehouses under possibility constraints. *Fuzzy Set Syst* 157(1):52–73.

Maiti, M. K., and M. Maiti. 2007a. Two-storage inventory model with lot-sizing dependent fuzzy lead-time under possibility constraints via genetic algorithm. *Eur J Oper Res* 179(2):352–371.

Maiti, M. K., and M. Maiti. 2007b. Two-storage inventory model in a mixed environment. *Fuzzy Optim Decis Making* 6(4):391–426.

Mandal, N. K., and T. K. Roy. 2006a. Multi-item imperfect production lot size model with hybrid number cost parameters. *Appl Math Comput* 182(2):1219–1230.

Mandal, N. K., and T. K. Roy. 2006b. A displayed inventory model with L-R fuzzy number. *Fuzzy Optim Decis Making* 5(3):227–243.

Mandal, N. K., T. K. Roy, and M. Maiti. 2005. Multi-objective fuzzy inventory model with three constraints: A geometric programming approach. *Fuzzy Set Syst* 150(1):87–106.

Mondal, S., and M. Maiti. 2002. Multi-item fuzzy EOQ models using genetic algorithm. *Comput Ind Eng* 44(1):105–117.

Ouyang, L.-Y., and H.-C. Chang. 2001. The variable lead time stochastic inventory model with a fuzzy backorder rate. *J Oper Res Soc Jpn* 44(1):19–33.

Ouyang, L.-H., and H.-C. Chang. 2002. A minimax distribution free procedure for mixed inventory models involving variable lead time with fuzzy lot sales. *Int. J Prod Econ* 76(1):1–12.

Ouyang, L.-Y., and J.-S. Yao. 2002. A minimax distribution free procedure for mixed inventory model involving variable lead time with fuzzy demand. *Comput Oper Res* 29(5): 471–487.

Ouyang, L.-Y., K.-S. Wu, and C.-H. Ho. 2006. Analysis of optimal vendor–buyer integrated inventory policy involving defective items. *Int J Adv Manuf Tech* 29(11/12):1232–1245.

Pai, P.-F., and M.-M. Hsu. 2003. Continuous review reorder point problems in a fuzzy environment. *Int J Adv Manuf Tech* 22(5/6):436–440.

Panda, D., S. Kar, K. Maity, and M. Maiti. 2008. A single period inventory model with imperfect production and stochastic demand under chance and imprecise constraints. *Eur J Oper Res* 188(1):121–139.

Pappis, C. P., and N. I. Karacapilidis. 1995. Lot size scheduling using fuzzy numbers. *Int Trans Oper Res* 2(2):205–212.

Park, K. S. 1987. Fuzzy-set theoretic interpretation of economic order quantity. *IEEE Trans Syst Man Cybern Syst Hum* 17(6):1082–1084.

Petrovic, D., R. Petrovic, and M. Vujosevic. 1996. Fuzzy models for the newsboy problem. *Int J Prod Econ* 45(1/3):435–441.

Petrovic, R., and D. Petrovic. 2001. Multicriteria ranking of inventory replenishment policies in the presence of uncertainty in customer demand. *Int J Prod Econ* 71(1/3):439–446.

Roy, A., S. Kar, and M. Maiti. 2008. A deteriorating multi-item inventory model with fuzzy costs and resources based on two different defuzzification techniques. *Appl Math Model* 32(2):208–223.

Roy, T. K., and M. Maiti. 1995. A fuzzy inventory model with constraint. *OPSEARCH* 32(4): 287–298.

Roy, T. K., and M. Maiti. 1997. A fuzzy EOQ model with demand-dependent unit cost under limited storage capacity. *Eur J Oper Res* 99(2):425–432.

Roy, T. K., and M. Maiti. 1998. Multi-objective inventory models of deteriorating items with some constraints in a fuzzy environment. *Comput Oper Res* 25(12):1085–1095.

Samanta, B., and S. A. Al-Araimi. 2001. An inventory control model using fuzzy logic. *Int J Prod Econ* 73(3):217–226.

Shao, Z., and X. Ji . 2006. Fuzzy multi-product constraint newsboy problem. *Appl Math Comput* 180(1):7–15.

Sinha, S., and S. P. Sarmah. 2008. An application of fuzzy set theory for supply chain coordination. *Int J Manag Sci Eng Manag* 3(1):19–32.

Sommer, G. 1981. Fuzzy inventory scheduling. In *Applied systems and cybernetics*, vol. VI, edited G. Lasker, 3052–3060. New York: Pergamon Press.

Tutuncu, G. Y., O. Akoz, A. Apaydin, and D. Petrovic. 2008. Continuous review inventory control in the presence of fuzzy costs. *Int J Prod Econ* 113(2):775–784.

Vijayan, T., and M. Kumaran. 2008. Inventory models with a mixture of backorders and lost sales under fuzzy cost. *Eur J Oper Res* 189(1):105–119.

Vujosevic, M., D. Petrovic, and R. Petrovic. 1996. EOQ formula when inventory cost is fuzzy. *Int J Prod Econ* 45(1/3):499–504.

Wang, X., W. Tang, and R. Zhao. 2007. Fuzzy economic order quantity inventory models without backordering. *Tsinghua Sci Tech* 12(1):91–96.

Wu, K., and J.-S. Yao. 2003. Fuzzy inventory with backorder for fuzzy order quantity and fuzzy shortage quantity. *Eur J Oper Res* 150(2):320–352.

Yadavalli, V. S. S., M. Jeeva, and R. Rajagopalan. 2005. Multi-item deterministic fuzzy inventory model. *Asia Pac J Oper Res* 22(3):287–295.

Yager, R. R. 1981. A procedure for ordering fuzzy subsets of the unit interval. *Inform Sci* 24: 143–161.

Yang, M. F. 2006. A two-echelon inventory model with fuzzy annual demand in a supply chain. *J Inform Optim Sci* 27(3):537–550.

Yang, M. F. 2007. Optimal strategy for the integrated buyer-vendor model fuzzy annual demand and fuzzy adjustable production rate. *J Appl Sci* 7(7):1025–1029.

Yao, J.-S., S.-C. Chang, and J.-S. Su. 2000. Fuzzy inventory without backordered for fuzzy order quantity and fuzzy total demand quantity. *Comput Oper Res* 27(10):935–962.

Yao, J.-S., M.-S. Chen, and H.-F. Lu. 2006. A fuzzy stochastic single-period model for cash management. *Eur J Oper Res* 170(1):72–90.

Yao, J.-S., and J. Chiang. 2003. Inventory without backorder with fuzzy total cost and fuzzy storing cost defuzzified by centroid and signed distance. *Eur J Oper Res* 148(2): 401–409.

Yao, J.-S., and H.-M. Lee. 1996. Fuzzy inventory with backorder for fuzzy order quantity. *Inform Sci* 93(3/4):283–319.

Yao, J.-S., and H.-M. Lee. 1999. Fuzzy inventory with or without backorder for fuzzy order quantity with trapezoid fuzzy number. *Fuzzy Set Syst* 105(3):311–337.

Yao, J.-S., L.-Y. Ouyang, and H.-C. Chang. 2003. Models for a fuzzy inventory of two replaceable merchandises without backorder based on the signed distance of fuzzy sets. *Eur J Oper Res* 150(3):601–616.

Yao, J.-S., and J.-S. Su. 2000. Fuzzy inventory with backorder for fuzzy total demand based on interval-valued fuzzy set. *Eur J Oper Res* 124(2):390–408.

Zadeh, L. A. 1965. Fuzzy sets. *Inform Contr* 8(3):338–353.

Nonfuzzy Literature

Aggarwal, S. C. 1974. A review of current inventory theory and its application. *Int J Prod Res* 12(4):443–482.

Anderson, E. T., G. J. Fitzsimons, and D. Simester. 2006. Measuring and mitigating the costs of stockouts. *Manag Sci* 52(11):1751–1763.

Arrow, K., T. Harris, and J. Marschak. 1951. Optimal inventory policy. *Econometrica* 19(3):250–272.

Badinelli, R. D. 1986. Optimal safety stock investment through subjective evaluation of stockout costs. *Decis Sci* 17(3):312–328.

Banerjee, A. 1986. A joint economic-lot-size model for purchaser and vendor. *Decis Sci* 17(3):292–311.

Ben-Daya, M., M. Darwish, and K. Ertogral. 2008. The joint economic lot sizing model: Review and extensions. *Eur J Oper Res* 185(2):726–742.

Benton, W. C., and S. Park. 1996. A classification of literature on determining the lot size under quantity discounts. *Eur J Oper Res* 92(2):219–238.

Berling, P. 2008. Holding cost determination: An activity-based cost approach. *Int J Prod Econ* 112(2):829–840.

Biggs, J. B., E. A. Thies, and J. R. Sisak. 1990. The cost of ordering. *Journal J Purch Mater Manag* 26(3):30–36.

Bock, R. H. 1964. Measuring cost parameters in inventory models. *Manag Tech* 4(1):59–66.

Bramson, M. J. 1962. The variable lead-time problem in inventory control: A survey of the literature, Part I. *Oper Res* 13(1):41–35.

Brown, R. M. 1985. On carrying costs and the EOQ model: A pedagogical note. *Financ Rev* 20(4):357–360.

Brown, R. M., T. E. Conine, and M. Tamarkin. 1986. A note of holding costs and lot-size errors. *Decis Sci* 17(4):603–610.

Buffa, E. S., and R. K. Sarin. 1987. *Production/operations management.* New York: Wiley & Sons.

Campbell, R. J., and L. J. Rankin. 1998. What customer orders really cost. *Mid Am J Bus* 13(1):41–47.

Chang, Y. S., and P. Niland. 1967. A model for measuring stock depletion costs. *Oper Res* 15(3):427–447.

Chitgopekar, S. S. 1990. Sensitivity analysis of the EOQ model with planned shortages. *OPSEARCH* 27(2):102–108.

Coopersmith, J. 1990. The cost of fulfillment. *Catalog Age* 7(11):111–114.

Denicoff, A. M., J. P. Fennell, and H. Solomon. 1960. Summary of a method for determining the military worth of spare parts. *Nav Res Logist Q* 7(3):221–234.

Dobson, G. 1988. Sensitivity of the EOQ model to parameter estimates. *Oper Res* 36(4):570–574.

Dolan, R. J. 1978. A normative model of industrial buyer response to quantity discounts. In *Research frontiers in marketing: Dialogues and directions*, edited by S. C. Jain, 121–125. Chicago: American Marketing Association.

Dworetzky, A., J. Kiefer, and J. Wolfowitz, 1952a. The inventory problem: I. Case of known distribution of demand. *Econometrica* 20(2):187–222.

Dworetzky, A., J. Kiefer, and J. Wolfowitz. 1952b. The inventory problem: II. Case of unknown distribution of demand. *Econometrica* 20(3):450–466.

Eilon, S. 1956. Economic lot sizes in batch production. *Engineering* 182(4729):522–523.

Gardner, E. S. 1980. Inventory management and the gods of Olympus. *Interfaces* 10(4):42–45.

Goyal, S. K., and B. C. Giri. 2001. Recent trends in modeling of deteriorating inventory. *Eur J Oper Res* 134(1):1–16.

Goyal, S. K., and F. Nebebe. 2000. Determination of economic production-shipment policy for a single-vendor–single-buyer system. *Eur J Oper Res* 121(1):175–178.

Guiffrida, A. L., and T. Papp. 2008. On the robust properties of the economic order quantity model. *Proc of the Int Acad of Bus and Econ*, Las Vegas, 20–22.

Harris, F. 1913. How many parts to make at once. *Factory Mag Manag* 10(2):136–152.

Jamal, A. M. M., B. B. Sarkar, and S. Wang. 1997. An ordering policy for deteriorating items with allowable shortage and permissible delay in payment. *J Oper Res Soc* 48(8):826–833.

Johnson, L., and D. Montgomery. 1974. *Operations research in production planning, scheduling and inventory control*. New York: Wiley & Sons.

Jordan, H. H. 2008. Calculating the true cost of inventory. *APICS- Perform Advantage* (February):58–61.

Karr, H. W. 1958. A method of estimating spare-part essentiality. *Nav Res Logist Q* 5(1):29–42.

Kennedy, W. J., J. W. Patterson, and L. D. Fredendall. 2002. An overview of recent literature on spare parts inventory. *Int J Prod Econ* 76(2):210–215.

Khouja, M. 1999. The single period (news-vendor) problem: Literature review and suggestions for future research. *Omega* 27(5):537–553.

Kim, S.-K. 1991. An analysis of inventory stockout models. PhD thesis, Pennsylvania State Univ.

Lambert, D. M. 1975. The development of an inventory costing methodology: A study of the costs associated with holding inventory. PhD thesis, Ohio State Univ.

Lambert, D., M. Bennion, and J. Taylor. 1990. The small order problem. *Manag Decis* 28(3):39–45.

Lowe, T. J. and L. B. Schwarz. 1983. Parameter estimation for the EOQ lot size model: Minimax and expected value choices. *Nav Res Logist Q* 30(2):367–376.

Malucci, L. J. 2008. What is the order cost. *APICS Magazine* (March/April):12.

Minner, S. 2003. Multiple-supplier inventory models in supply chain management: A review. *Int J Prod Econ* 81–82:265–279.

Moon, I., and S. Choi. 1998. A note on lead time and distributional assumptions in continuous review inventory models. *Comput Oper Res* 25(11):1007–1012.

Nahmias, S. 1978. Inventory models. In *Encyclopedia of computer science and technology*, edited by J. Belzer, A. G. Hlzman, and A. Kent, 447–483. New York: Marcel Dekker.

Nahmias, S. 1982. Perishable inventory theory: A review. *Oper Res* 30(4):680–708.

Nose, T., H. Ishii, and T. Nishida. 1981. Some properties of perishable inventory control subject to stochastic lead-time. *J Oper Res Soc Jpn* 24:110–135.

Oral, M., M. S. Salvador, A. Reisman, and B.V. Dean. 1972. On the evaluation of shortage costs for inventory control of finished goods. *Manag Sci* 18(6):B-344–B-351.

Ouyang, L. Y., and B. R. Chuang. 2001. A periodic review inventory-control system with variable lead time. *Int J Inform Manag Sci* 12(1):1–13.

Ouyang, L. Y., N. C. Yeh, and K. S. Wu. 1996. Mixture inventory model with backorders and lost sales for variable lead time. *J Oper Res Soc* 47(6):829–832.

Peterson, R., and E. Silver. 1979. *Decision systems for inventory management and production planning*. New York: Wiley.

Porteus, E. L. 1986. Optimal lot sizing, process quality improvement and set up cost reduction. *Oper Res* 34(1):137–144.

Raafat, F. 1991. Survey of literature on continuously deteriorating inventory models. *J Oper Res Soc* 42(1):27–37.

Reid, R. A., C. Huth, and D. N. Bryson. 1984. Inventory cost determination: A public sector challenge. *J Purch Mater Manag* 20(4):27–31.

Reuter, V. C. 1978. The big gap in inventory management. *J Purch Mater Manag* 14(3): 17–22.

Salameh, M. K., and M. Y. Jaber. 2000. Economic production quantity model for items with imperfect quality. *Int J Prod Econ* 64(1–3):59–64.

Schary, P. B., and M. Christopher. 1979. The anatomy of a stock-out. *J Retailing* 55(2):59–70.

Schwartz, B. L. 1966. A new approach to stockout penalties. *Manag Sci* 12(12): B-538– B-544.

Silver, E. A. 1981. Operations research in inventory management: A review and critique. *Oper Res* 29(4):628–645.

Silver, E. A. 2008. Inventory management: An overview, Canadian publications, practical applications and suggestions for future research. *INFOR* 46(1):15–28.

Silver, E. A., D. F. Pyke, and R. Peterson. 1998. *Inventory management and production planning and scheduling*. New York: Wiley.

Sinha, S., and S. P. Sarmah. 2007. Buyer–vendor coordination through quantity discount policy under asymmetric cost information. *Proc IEEE Int Conf Ind Eng and Eng Manag*, 1558–1562.

Starr, M. K., and D. W. Miller. 1962. *Inventory control and practice*. Englewood Cliffs, NJ: Prentice-Hall.

Urban, T. 2005. Inventory models with inventory-level-dependent demand: A comprehensive review and unifying theory. *Eur J Oper Res* 162(3):792–804.

Vlachos, D., and R. Dekker. 2003. Return handling options and order quantities for single period products. *Eur J Oper Res* 151(1):38–52.

Wagner, H. M. 1969. *Principles of operations research, with applications to managerial decisions*. Englewood Cliffs, NJ: Prentice-Hall.

Wagner, H. M. 1980. Research portfolio for inventory management and production planning systems. *Oper Res* 28(3):445–475.

Walter, C. K., and J. R. Grabner. 1975. Stockout cost models: Empirical tests in a retail situation. *J Market* 39(3):56–68.

Williams, B. D., and T. Tokar. 2008. A review of inventory management research in major logistics journals. *Int J Logist Manag* 19(2):212–232.

Yang, P., and H. Wee. 2000. Economic ordering policy of deteriorated item for vendor and buyer: An integrated approach. *Prod Plann Contr* 11(5):474–480.

Yu, G. 1997. Robust economic order quantity models. *Eur J Oper Res* 100(3):482–493.

Zanakis, S. H., L. M. Austin, D. C. Nowading, and E. A. Silver. 1980. From teaching to implementing inventory management: Problems of translation. *Interfaces* 10(6):103–110.

Zangwill, W. I. 1987. From EOQ towards ZI. *Manag Sci* 33(10):1209–1221.

9 Modeling Hidden Costs of Inventory Systems
A Thermodynamic Approach

Mohamad Y. Jaber
Ryerson University
Toronto, Ontario

CONTENTS

9.1 INTRODUCTION

The earliest inventory model that is available in the literature is the economic order quantity (EOQ) model developed by Harris (1913). Since its introduction, Harris's model has been the basis for many models in inventory management, supply chain management, and reverse logistics. The popularity of the EOQ model has been recognized by both academics and practitioners (e.g., Osteryoung et al. 1986). Some of the positive features attributed to the popularity of the EOQ model are its simple mathematics and the insensitivity of the total cost to changes in the order quantity (batch size). Despite its popularity, the EOQ model has been criticized by some researchers on the basis that its assumptions are never met (e.g., Woolsey 1988). Some of these works are reviewed next.

Adkins (1984) observed that the application of EOQ in some firms led to results that ranged from disappointing to disastrous. Selen and Wood (1987) cautioned that substantial miscalculations or misinterpretations during parameter input (most specifically the holding cost and the order cost) determination often led to poor results. Woolsey (1988) severely critiqued using the EOQ model to the extent that he wrote: "If you continue to love and use the EOQ without knowing what it is costing you, I can only suggest that you deserve each other" (p. 72). Jones (1991) cautioned that most accountants fail to identify relevant costs in a system and, consequently, most manufacturers who use the EOQ formula to minimize relevant annual costs will calculate a lot size higher than what it would be if the model were employed correctly.

Therefore, the common cost parameters used in the EOQ model are aggregated costs, particularly the ordering cost, holding cost, and shortage cost, and some of their components are difficult to estimate, so the results obtained may be misleading. For example, the holding cost may include all (or some) of the following: cost of money tied up that is either borrowed (on which interest is paid) or could be put to other use (in which case there are opportunity costs); storage cost (supplying a warehouse, rent rates, heat, light, etc.); loss (due to damage, pilferage, and obsolescence); handling (including all movement, special packaging, refrigeration, putting on pallets, etc.); administration (stock checks, computer updates, etc.); insurance; and taxes (Waters 2003). In addition to the usual reported costs, there are hidden costs associated with production/inventory systems that are not usually accounted for (Ullmann 1982; Pendlebury and Platford 1988; Gooley 1995; Crusoe et al. 1999; Konar and Cohen 2001; Fisher and Siburg 2003; Callioni et al. 2005).

In recent years, there has been a call by a few researchers in the field to think outside the box of classical inventory management (e.g., Sprague 2002; Bonney et al. 2003; Chikán 2007). This may result in researchers attempting to bridge their fields with others to gain insight into their own, benefiting from the synergies of such processes. For example, some researchers in the discipline of operations research management science have applied Shannon's (1948) information theory and entropy approaches to account for disorder when modeling the behavior of production or supply systems (e.g., Karp and Ronen 1992; Ronen and Karp 1994; McCarthy et al. 2000). However, few have applied classical thermodynamics reasoning to modeling such systems. For example, Jaber et al. (2004) postulated that the behavior of production systems very much resembles that of physical systems. Such a parallel suggests that improvements to production systems may be achievable by applying the first and second laws of thermodynamics to reduce system entropy (or disorder). Jaber et al. (2004) introduced the concept of entropy cost to account for hidden and difficult-to-estimate costs associated with inventory systems. This chapter will discuss the works along this line of research, and will propose a future research direction.

The remainder of this chapter is organized as follows. Section 9.2 provides a review of the literature for those works that bridged thermodynamics and operations research/management science. Section 9.3 provides a background to the principles of the thermodynamics of commodity flow systems. Section 9.4 provides the mathematics for applying the laws of thermodynamics to the economic order quantity model. Section 9.5 discusses some of the recent extensions and applications of the model presented in Section 9.4. Section 9.6 extends the basic model presented in Section 9.4 by modeling the commodity flow function (or demand) to be dependent on price and quality rather than on price alone as in Section 9.4. Section 9.7 summarizes and concludes this chapter.

9.2   THERMODYNAMICS AND OPERATIONS RESEARCH

Operations research (OR) has always been synonymous with management science, with the names being interchangeable (e.g., Barish 1963). OR tools, such as optimization, queuing theory, game theory, decision analysis, and simulation, are being applied to solve real-world problems in inventory theory, supply chain and logistics management, production planning, scheduling, and so forth.

Researchers have been using OR tools to analyze and design more efficient and productive systems. However, few researchers started thinking of such systems as being analogous to physical systems (Goodeve 1953; Wilson 1970), while others attempted to bridge the two disciplines. Drechsler (1965) was the first who attempted to show how the "law of disorder" or the "second law of thermodynamics" (or entropy) influenced management thinking. The second law of thermodynamics suggests that every system, left unto itself, tends toward disorder (Drechsler 1968), which frequently appears in the form of queues of materials waiting to be processed, machines waiting for service, finished products waiting for customers, and so forth. Drechsler (1968) applied the law of disorder to analyze complex-decision-type systems where he modeled a decision tree similar to a steam-power generation plant using a thermodynamic analogy. Samaras (1973) argued that the effects of entropy include common problems of inefficiency and related elements such as higher costs, lower outputs, greater wastes, and lower profits. Entropy effects also include a lowered capability to adapt to changes in the world, low morale, and loss of motivation and business failure. Tyler (1989) presented a model of manpower systems and suggested that entropy can be evaluated from the tenure profiles of manpower systems.

This research interest, although slow, continued in recent years. Whewell (1997) argued that logistical processes for supply chains are as significant as are the laws of thermodynamics for technologists. By making parallels between a logistic system and a thermodynamic system, his analysis suggests that suppliers must give their customers increased choice, flexibility, and value from the service they provide if they want to win a larger market share from competitors. Harris (1999) argued that the second law of thermodynamics works. It helps understand the need to achieve higher availability; that is, to keep more customers coming back by having better service on less-costly items. Chen (1999) matched elements of the business process to components of thermodynamics, and compared the managerial functions of a socioeconomic system to the thermodynamic behavior of the physical universe. Two laws of the business process based on those of thermodynamics were derived. The first law states the conservation of resources. Losses in the process, which consume resources and erode system productivity, have been discussed. The second law discusses the governing principle that makes the output of a system acceptable to customers. Tseng

(2004) applied the first and second laws of thermodynamics to constrain the processes by which raw materials are transformed into consumable goods and these goods are distributed afterward. He suggested that the size of the sales force can be determined by introducing the concept of temperature from thermodynamics to represent the flow (demand) of goods as hot (high) or cold (low).

More recently, Jaber et al. (2004) postulated that the behavior of production systems very much resembles that of physical systems. Such a parallel suggests that improvements to production systems may be achievable by applying the first and second laws of thermodynamics to reduce system entropy (or disorder). The EOQ and just-in-time (JIT) models were used to demonstrate the applicability of these laws. Along the same line of research, several models have been developed by the authors in inventory management, supply chain management, and reverse logistics. These works will be discussed in Section 9.5.

9.3   THE LAWS OF THERMODYNAMICS AND THE CONCEPT OF ENTROPY

Thermodynamics is a branch of physics that deals with the energy and work of a system, the transfer of energy between the system and its surrounding, and the conversion of energy from one form to another. Temperature, volume, pressure, and chemical composition are the properties that define a thermodynamic system. A thermodynamic system is said to be in equilibrium when the values for these properties remain unchanged over time. Mainly, there are two laws (the first and the second) of thermodynamics that govern the creation of energy and its flow that are of interest here. For more details, the textbook on thermodynamics by Cengel and Boles (2002) is recommended.

9.3.1   BASIS FOR THE MODEL

The first law of thermodynamics states that energy is conserved, that is, energy cannot be created or destroyed. Energy can be added to a thermodynamic system from a high-temperature reservoir or lost to a low-temperature sink by a heat (work) process. The internal energy of a thermodynamic system is the amount added by heating (δQ) minus the amount lost by doing work (δW) on the system. This difference, $dU = \delta Q - \delta W$, is the internal energy of a thermodynamic system that is associated with the random, disordered motion of molecules. Analogously, if heat were money, then one could say that a change in savings (dU) is equal to the money deposited (δQ) minus the money withdrawn or spent (δW).

The second law of thermodynamics places constraints upon the direction of the flow of heat between the system and its surrounding. It suggests that it is not possible for a device to operate in a cycle receiving heat and producing a net equivalent amount of work; and that it is not possible for a device to operate in a cycle and produce no effect other than the transfer of heat from a lower-temperature reservoir to a higher-temperature reservoir. The second law of thermodynamics is always associated with the concept of entropy, which is a measure of the amount of energy that is unavailable to do work (wasted energy) or a measure of the disorder of a system.

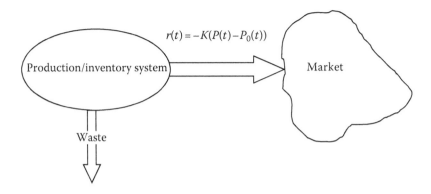

FIGURE 9.1 Flow of commodity from the system to the market.

The second law of thermodynamics also suggests that the entropy tends to increase with time, that is, entropy is an arrow of time itself. The change in entropy over time, dS/dt, is represented as the heat added or removed over time, dQ/dt, divided by the temperature, that is, $dS/dt = (dQ/dt)/T$ or $dS = dQ/T$.

A thermodynamic system is defined by its properties, which are temperature, volume, pressure, and chemical composition. Analogous to a thermodynamic system, a production system could be described by its characteristics too, which are price and quality of a commodity. For example, reducing (increasing) the price (temperature) of a commodity (energy) below (above) the market price (surrounding temperature) conceptually increases customers' demand (energy flows from a high- to a low-temperature reservoir), and produces a commodity flow (sales) from the system to its surrounding (market). Part of the work done on the system to generate this flow of commodity is wasted in accordance with the second law of thermodynamics. Figure 9.1 shows some of the commodity flows (sales) to the market that are converted to revenue (equivalent to useful work) and wasted work (hidden cost) that result from this flow process.

9.3.2 COMMODITY FLOW FUNCTION

When work is done on a thermodynamic system, the system goes through a change in its state (e.g., increase in temperature), and a certain flow ensues to the system's surrounding (which is at a lower temperature). Analogously, in a production system the price is reduced below the market equilibrium value ensuing a flow of the commodity to the market. The commodity flux, or demand rate ($r(t)$), is synonymous to the energy transferred, $\dot{Q} = dQ/dt = C(T_s - T)$, in a thermodynamic system, where C is the thermal capacity, T_s is the system temperature, and T is the temperature of the surrounding. The demand rate, $r(t)$, is then given as

$$r(t) = -K(P(t) - P_0(t)) \tag{9.1}$$

where $P_0(t)$ is the market equilibrium price at time t, $P(t)$ is the unit price at time t, and K (analogous to a thermal capacity) is the change in the flux for a change in the price $(P(t) - P_0(t))$ of a commodity, and is measured in additional units per year per change in unit price, e.g., units/year/\$. Equation 9.1 is also similar in form to excess demand functions discussed in mathematical economics literature (e.g., Arrow and Debreu 1954; Uzawa 1960).

The operations management literature (which includes operations research/ management science) advocates that a firm wishing to acquire a market share for its product may follow one of the following strategies: (1) the firm provides the same quality product as its competitor at a lower price, (2) the firm provides a better quality product than its competitor for the same price, or (3) perhaps, a firm adopts an aggressive policy by providing a better quality product than its competitor at a lower price.

Jaber et al. (2004) adopted the first strategy as described in Equation 9.1, where $P(t) < P_0(t)$ for every $t \in [0,T]$, where T is the inventory cycle time; a strategy of increasing commodity flow. With this strategy, the equilibrium price $P_0(t) - P_0$ is a constant for every $t \in [0,T]$, and $P(t)$ is a monotonically decreasing function over the specified interval, that is, $P'(t) < 0$ and $P''(t) > 0$ for every $t \in [0,T]$. This strategy is assumed to increase customers' demand, that is, $r(t)$ increases over the interval $[0,T]$ as a result of the price discounts $(P(t) - P_0(t) < 0)$.

9.3.3 ENTROPY AND ITS COST

Heat, or energy, can never pass spontaneously from a cold to a hot body. Work needs to be done on the system to allow energy to transfer. This transfer has one direction as the process is irreversible. A parallel could be drawn for a supply system. For example, a supplier that seeks to increase its share of the market will attempt to increase and sustain customer satisfaction relative to its competitors, and/or attempt to reduce costs associated with providing the same level of satisfaction (Whewell 1997, 18). Whewell (1997) advocates that changes to improve customer satisfaction can alter the established practices in at least one part of the supply process. Such changes can create disorder (entropy) and the costs associated with it are difficult to quantify.

Note that when $P(t) - P_0(t)$, the direction of the commodity flow is from the system to its surrounding, that is, from a high-temperature reservoir (lower price) to a lower-temperature reservoir (higher market equilibrium price). The entropy generation rate must satisfy (Cengel and Boles 2002)

$$S(t) = \frac{d\sigma(t)}{dt} = r(t)\left(-\frac{1}{P_0(t)} + \frac{1}{P(t)}\right) = K\left(\frac{P(t)}{P_0(t)} + \frac{P_0(t)}{P(t)} - 2\right) \qquad (9.2)$$

where $r(t)$ is given from Equation 9.1, $\sigma(t)$ is the total entropy generated by time t, and $S(t)$ is the rate at which entropy is generated over time. The total demand in a cycle of duration T, $D(T)$, is taken from Equation 9.1 as

$$D(T) = \int_0^T r(t)dt = \int_0^T -K(P(t) - P_0(t))dt \qquad (9.3)$$

where $D(T)$ is measured in units and it is the lot size quantity, that is, $Q = D(T)$, if all units are of good quality and if shortages do not occur. Similarly, the total entropy generated in a cycle of duration T, $\sigma(T)$, is taken from Equation 9.2 as

$$\sigma(T) = \int_0^T S(t)dt = \int_0^T K\left(\frac{P(t)}{P_0(t)} + \frac{P_0(t)}{P(t)} - 2\right)dt \qquad (9.4)$$

Then the entropy cost per cycle, $E(T)$, is given by dividing the expression in Equation 9.3 by that in Equation 9.4 to get

$$E(T) = \frac{D(T)}{\sigma(T)} = \frac{\int_0^T -K(P(t) - P_0(t))dt}{K\int_0^T \left(\frac{P(t)}{P_0(t)} + \frac{P_0(t)}{P(t)} - 2\right)dt} \qquad (9.5)$$

where $E(T)$ is measured in dollars.

9.4 A PRICE-DRIVEN ECONOMIC ORDER QUANTITY MODEL

A typical inventory model will relate the total cost of operating the system to surplus (holding) cost, shortage (lost sales or back order) cost, and initiation (i.e., ordering, setup, and planning) cost (e.g., Bonney 1994). The EOQ model cost function is the sum of two conflicting costs, which are the holding, $HC(T)$, and the order, A, costs per cycle of length T. In the EOQ model, the stock of a commodity is replenished every T units of time by Q units. These Q units are depleted at a rate of $r(T)$, with the inventory level at time t, $I(t)$, and is taken from Equations 9.1 and 9.3 as

$$I(t) = Q - \int_0^t r(t)dt = D(T) - \int_0^t -K(P(y) - P_0(y))dy \qquad (9.6)$$

Note that $I(0) = D(T)$ and $I(T) = 0$, where $t \in [0,T]$. The total cost per cycle for an EOQ model is given as

$$C(T) = A + HC(T) = A + h\int_0^T I(t)dt = A + h\int_0^T \left(D(T) - \int_0^t -K(P(y) - P_0(y))dy\right)dt \qquad (9.7)$$

where h is the holding cost per unit per unit of time. The unit time cost function is given by dividing Equation 9.7 by T to get

$$c(T) = \frac{C(T)}{T} = \frac{A + HC(T)}{T} = \frac{A}{T} + \frac{h}{T}\int_0^T \left(D(T) - \int_0^t -K(P(y) - P_0(y))dy\right)dt \qquad (9.8)$$

As discussed earlier, there is no doubt that the proper estimation of the EOQ model input parameters—order cost, carrying (holding) cost, and the demand rate—is essential for producing reliable results. To address this problem, Jaber et al. (2004) proposed accounting for an additional cost, the entropy cost in Equation 9.5, when analyzing EOQ systems. Therefore, Equation 9.8, after adding Equation 9.5, is written as

$$\psi(T) = \frac{c(T) + E(T)}{T} = \frac{C(T)}{T} = \frac{A + HC(T) + E(T)}{T}$$

$$= \frac{A}{T} + \frac{h}{T}\int_0^T \left(\int_0^T -K(P(t) - P_0(t))dt - \int_0^t -K(P(y) - P_0(y))dy\right)dt$$

$$+ \frac{\int_0^T -K(P(t) - P_0(t))dt}{TK\int_0^T \left(\frac{P(t)}{P_0(t)} + \frac{P_0(t)}{P(t)} - 2\right)dt} \tag{9.9}$$

The EOQ model assumes a constant demand rate, that is, $r = r(t)$ where $t \in [0, \infty)$, which implies that Equation 9.1 reduces to $r = -K(P - P_0)$ such that $P < P_0$. This assumption reduces Equation 9.8 to

$$c(T) = \frac{A}{T} + hr\frac{T}{2} \tag{9.10}$$

whose optimal solution is $T^* = \sqrt{2A/hr}$, or equivalently $Q^* = \sqrt{2Ar/h}$, which are the optimal cycle time and the optimal order quantity (lot size), respectively. Similarly, Equation 9.5 reduces to

$$E(T) = \frac{D(T)}{\sigma(T)} = \frac{\int_0^T -K(P - P_0)dt}{K\int_0^T \left(\frac{P}{P_0} + \frac{P_0}{P} - 2\right)dt} = -\frac{PP_0}{P - P_0} \tag{9.11}$$

Then Equation 9.9 is written from Equations 9.10 and 9.11 as

$$\psi(T) = \frac{A}{T} + hr\frac{T}{2} - \frac{PP_0}{(P - P_0)T} = \frac{A}{T} + hr\frac{T}{2} + \frac{E}{T} \tag{9.12}$$

where $E = P_0P/(P_0 - P)$ and $r = -K(P - P_0)$. Equation 9.12 is convex since $d^2\psi(T)/dT^2 > 0 \ \forall \ T > 0$. Setting $d\psi(T)/dT = 0$ and solving for T results in

$$T^{**} = \sqrt{\frac{2(A + E)}{hr}} \tag{9.13a}$$

and

$$Q^{**} = \sqrt{\frac{2r(A + E)}{h}} \tag{9.13b}$$

Substituting Equation 9.13a in Equation 9.12, Equation 9.12 reduces to

$$\psi^{**} = \sqrt{2hr(A+E)}$$
(9.14)

Taking the ratio of $T^{**}/T^* = Q^{**}/Q^*$ results in

$$\frac{Q^{**}}{Q^*} = \sqrt{1 + \frac{E}{A}}$$
(9.15)

The cost of controlling the flow of one unit of commodity is determined by dividing Equation 9.5 by Equation 9.13b to get

$$u = \frac{E}{Q^*} = \frac{E}{\sqrt{\frac{2r(A+E)}{h}}}$$
(9.16)

From Equations 9.15 and 9.16, one can realize that as P and P_0 decrease while keeping $P_0 - P$ and A constant, and $E = P_0 P/(P_0 - P)$ decreases, $Q^{**}/Q^* \to 1$ and $u \to 0$ when P and P_0 become very small. Furthermore, decreasing P and P_0, while increasing $P - P_0$, suggests that E approaches a very low value, making it insignificant. These observations suggest that it is more relevant to account for entropy cost for expensive and low-demand items rather than inexpensive and high-demand items.

As markets became more dynamic and global, manufacturing firms have been faced with fierce competition and a rapidly changing environment. This led many manufacturing firms to introduce and implement various improvement programs to enhance their competitiveness. Among the managerial philosophies invented and implemented is just-in-time (JIT), which generally seeks the elimination of waste in all aspects of a firm's production activities: human relations, vendor relations, technology, and management of materials (Chen 1999). JIT promotes smaller lots of materials or products for flexibility. Although smaller batch sizes provide greater flexibility, they may increase the control costs (entropy cost). Some researchers have echoed this implicitly. Ullmann (1982) identified excess managerial cost to control the improvement process as a hidden cost. Crawford et al. (1988) identified costly problems (e.g., accounting, lack of training, interference between new and old methods) encountered by JIT implementers. Cavinato (1991) cautioned that smaller-lot size policies require tighter management and monitoring of the logistics information system in real-time, which is costly. Lummus et al. (1995) identified "confusion" in pricing of customized products as a hidden cost. Crusoe et al. (1999) listed increased labor union leverage, the inappropriateness of flexible manufacturing systems (FMSs) for certain products, the difficulties in maintaining a steady stream of incoming materials when dealing with commodities, and the increased space requirements due to the rearrangement of machines into cells, as some of the hidden costs of JIT systems.

JIT turns the EOQ formula around. Instead of accepting setup times as fixed, companies work to reduce setup time and reduce lot sizes. Ideally, the optimal order quantity is equal to the daily demand, that is $Q_j^{**} = r$. Then, from Equation 9.13b

$r = \sqrt{2r(A+E)/h}$, and the optimal setup is $A_J = rh/2 - E$. Since $A_J = rh/2 - E > 0$, this implies that the holding cost is $h > 2E/r$ or $h > 2P_0 P/(P_0 - P)^2$. In a JIT system, and from Equation 9.16, the cost of controlling the flow of one unit of commodity is $u_J = E/r$, where $Q_J^{**} = r$. This implies that it is more expensive to control the flow of commodity in a JIT system than in a classical EOQ system, where $u_J > u$ since $Q_J^{**} < Q^{**}$. This may explain the hidden costs indicated in some of the literature.

9.5 EXTENSIONS AND APPLICATIONS

The model developed by Jaber et al. (2004), described in Equation 9.9, was extended to other inventory situations and its applicability was investigated in supply chain and reverse logistics contexts. The extensions to the model of Jaber et al. (2004) and its applications are briefly discussed in the following subsections.

9.5.1 INVENTORY MANAGEMENT

The work of Jaber et al. (2006) differs from the work of Jaber et al. (2004) by assuming a constant rather than an increasing demand rate for a finite rather than an infinite planning horizon. In Jaber et al. (2006) the price of the commodity continues to decrease as a result of competition in the market, where the firm can predict with certainty that the equilibrium price would reach a target value $P_0(\tau)$, where $P_0(0) > P_0(\tau)$, and τ is some future point in time when the firm has no information on how the market price will behave. Their results suggested that for large values of the order cost, A, it might not be necessary to include the entropy cost. This may caution managers who wish to reduce the order (setup) costs to simultaneously reduce entropy costs.

Jaber (2007) extended the model of Jaber et al. (2004) to account for permissible delays in payments. Like quantity discounts (e.g., Crowther 1964; Dolan 1987), permissible delays in payments is a business practice used by many suppliers to promote commodities and gain larger market shares by offering credit terms to their customers. The EOQ model was investigated for permissible delays in payments (e.g., Haley and Higgins 1973; Kingsman 1983; Goyal 1985; Salameh et al. 2003). From Jaber (2007), Equation 9.12 is rewritten as

$$\psi(T) = \frac{A}{T} + \frac{hr}{2}\frac{(T-y)^2}{T} + \frac{E}{T} - \frac{rkP}{2T}y^2 \tag{9.17}$$

where T is the cycle time, y is the permissible delay in payment and $y \leq T$, k is the investment rate, and $h > kP$. The solution for Equation 9.17 is given by setting the first derivative of Equation 9.17 equal to zero and solving for T to get

$$T^* = \sqrt{\frac{2(A+E)+(h-kP)ry^2}{hr}} \tag{9.18a}$$

and

$$Q^* = \sqrt{\frac{2r(A+E)+(h-kP)(ry)^2}{h}}$$ (9.18b)

Equation 9.18b is identical to that in Goyal (1985) except for the term $E(T) = E = P_0 P/(P_0 - P)$, where the ratio of Q^*/Q^*_{Goyal} reduces to

$$\frac{Q^*}{Q^*_{Goyal}} = \sqrt{1 + \frac{2rE}{2rA+(h-kP)(ry)^2}}$$ (9.19)

Equation 9.19 indicates that as demand, r, and order cost, A, approach large values, then $Q^*/Q^*_{Goyal} \to 1$. As P and P_0 decrease while keeping $P_0 - P$ constant, $E = P_0 P/(P_0 - P)$ decreases, suggesting that $Q^*/Q^*_{Goyal} \to 1$ when P and P_0 are very small. These observations also suggest that accounting for entropy costs may be more relevant for low-demand and expensive items. When there is no delay in payments, $y = 0$, Equations 9.17, 9.18a, 9.18b, and 9.19 reduce to Equations 9.12, 9.13a, 9.13b, and 9.15, respectively. With delays in payments the order quantity is larger, that is, Equation 9.18b is greater than Equation 9.13b; then from Equation 9.16, the cost of controlling the flow of one unit of commodity becomes cheaper.

Jaber et al. (2009) investigated the model of Jaber et al. (2004) for deterioration effects. The effect of deteriorating items on inventory policies has been studied extensively, and review papers of inventory models for deteriorating items have been written by Rafaat (1991) and Goyal and Giri (2003); the EOQ model has been considered in the largest share of these studies. Jaber et al.'s (2009) objective was to examine whether the effect of deterioration counterbalances the earlier suggested increases in batch sizes and leads to smaller batches. Here, we present a simplified form of the model presented in Jaber et al. (2009), then Equation 9.12 can be rewritten as

$$\psi(T) = \frac{A}{T} + \frac{E[(1-\alpha)^{-T}-1]}{T\ln(1-\alpha)} + rh\frac{[-1+(1-\alpha)^{-T}+T\ln(1-\alpha)]}{T\ln(1-\alpha)^2}$$ (9.20)

Equation 9.20 can be solved using a numerical search technique.

Jaber et al. (2009a) extended the classical economic manufacture (order) quantity (EMQ/EOQ) model and the lot-sizing problem with learning and forgetting (Jaber and Bonney 1998) by including the entropy cost concept introduced in Jaber et al. (2004).

Jaber et al. (2009b) applied the concept of entropy cost to extend the classical EOQ model under the assumptions of perfect and imperfect quality. They did so by investigating the model of Porteus (1986) for the concept of entropy (Jaber et al. 2004). Porteus (1986) modified the EOQ model by assuming that when producing a single unit of the product, the production process (machine) shifts to an out-of-control state with a Markov transition probability (ρ) and the production process begins to produce defective products. Once the process is out of control it remains in that state until the entire lot of size Q is produced. The total cost function was given as

$$\psi(T) = \frac{A}{T} + c_u r + h\left(1 + \rho\frac{c_R}{2c_u}\right)\frac{rT}{2} + c_R \rho r^2 \frac{T}{2} + \frac{E}{T} \tag{9.21}$$

The optimal solution is given by setting the first derivative of Equation 9.21 to zero to get

$$T^* = \sqrt{\frac{2(A+E)}{h(1 + \rho c_R/2c_u)r + c_R \rho r^2}} \tag{9.22a}$$

and

$$Q^* = \sqrt{\frac{2r(A+E)}{h(1 + \rho c_R/2c_u) + c_R \rho r}} \tag{9.22b}$$

where c_u is the unit cost; c_R is the cost of reworking a defective unit; and E, r, A, and h are as defined earlier in this section. Note that when $\rho = 0$, Equations 9.21, 9.22a, and 9.22b reduce to Equations 9.12, 9.13a, and 9.13b, respectively.

9.5.2 SUPPLY CHAIN MANAGEMENT

Jaber et al. (2006) applied the concept of entropy cost developed by Jaber et al. (2004) to a two-level (vendor–buyer; e.g., Goyal 1977; Banerjee 1986) supply chain, where they assumed a finite planning horizon. In their model, a vendor's cycle has λ buyer's cycles, where λ is a positive integer. If the buyer orders n times a year, then the vendor has m replenishment cycles a year, where $m\lambda = n$ and m and n are positive integers. Since the vendor's demand flow is a stepped function, the vendor's commodity flow is written as

$$r_v(t) = -K_v(P_v(t) - P_{0,v}(t)) \tag{9.23}$$

where $P_v(t)$ and $P_{0v}(t)$ are respectively the vendor's commodity and equilibrium price functions, and K_v is similar in definition to K (later referred to by K_b where the subscript b indicates the buyer). The entropy generated in the jth vendor's cycle that has λ buyer's cycles was given as

$$\sigma_{v,j}(\lambda,n) = \sum_{i=1}^{\lambda} K_v \left\{ \frac{P_v(t)}{P_{0,v}(t)} + \frac{P_{0,v}(t)}{P_v(t)} - 2 \right\}$$

$$= \sum_{i=1}^{\lambda} K_v \left\{ \frac{P_v(i,j,\lambda,n)}{P_{0,v}(i,j,\lambda,n)} + \frac{P_{0,v}(i,j,\lambda,n)}{P_v(i,j,\lambda,n)} - 2 \right\} \tag{9.24}$$

where $i = 1, 2, \dots, \lambda$, $j = 1, 2, \dots, m$, and $T = \tau/n$, with τ being the length of the planning horizon. The entropy cost of the jth vendor's cycle was given as

$$E_{v,j}(\lambda, n) = \frac{Q\lambda}{\sigma_{v,j}(\lambda, n)} \qquad (9.25)$$

The total supply chain cost was given as

$$T_{sc}(\lambda, n) = nA_b + h_b \frac{D\tau^2}{2n} + \sum_{j=1}^{m} E_b(\lambda, n) + \frac{n}{\lambda} A_v + \frac{h_v}{2}(\lambda - 1)D\frac{\tau^2}{n} + \sum_{j=1}^{n/\lambda} E_{v,j}(\lambda, n) \qquad (9.26)$$

where A_b is the buyer's order cost, h_b is the buyer's holding cost per unit per unit time, E_b is the entropy cost per buyer's cycle, and r is the demand rate per unit time and assumed to be constant and uniform over time, A_v is the vendor's order (setup) cost, h_v is the vendor's holding cost per unit per unit time, and λ is the vendor lot-size multiplier (positive integer) of the buyer's order quantity Q.

9.5.3 REVERSE LOGISTICS

Since the 1980s, manufacturers have been moving toward shorter product life cycles, smaller lot sizes (e.g., JIT), and frequent deliveries to sustain their competitive advantage in a dynamic and global marketplace. This quickened the forward flow of products in a supply chain resulting in faster rates of product waste generation and depletion of natural resources. Consequently, manufacturing and production processes have been viewed as culprits in harming the environment (Fiksel 1996, in Beamon 1999). This concern gave rise to the concept of reverse logistics (RL) or the backward flow of products from customers to manufacturers to suppliers (e.g., Gungor and Gupta 1999). Like supply chains, reverse logistics manages the flow of products to be recovered, however, in the opposite direction (i.e., from downstream to upstream), and therefore managing inventory in reverse logistics is of importance (e.g., Fleischmann et al. 1997).

Jober and Rosen (2008) applied the first and second laws of thermodynamics and the concept of entropy cost developed by Jaber et al. (2004) in a closed-loop model (Richter 1996a, 1996b). The production environment described in Richter (1996a, 1996b) consists of two shops. The first shop is for production and recovery (e.g., repair, remanufacturing), while the second shop is for collecting used/returned items from the market at some rate. Richter (1996a, 1996b) investigated the EOQ model where recovered items are considered as good as new. The total cost function was an aggregated cost of production, recovery, and holding costs of produced/recovered and used items. Since this model assumes both forward and backward flows of goods to and from the market, Figure 9.1 is altered to model the collection rate (commodity flow in the reverse direction), where $V(t)$ is the collection price the manufacturer is willing to pay for a used product, and $V_0(t)$ is the collection price a customer can obtain for his/her used product if sold individually where $P_0(t) > P(t) > V(t) > V_0(t)$.

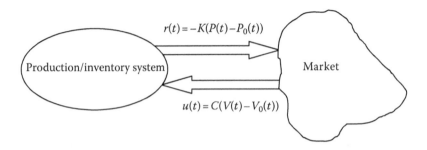

FIGURE 9.2 A cyclic commodity flows between the production/inventory system and its market.

Like Equation 9.1, the collection flux of used products, or the recovery rate $u(t)$, can be written as (Jaber and Rosen 2008)

$$u(t) = C(V(t) - V_0(t)) \tag{9.27}$$

where C is similar in definition to K in Equation 9.1. Figure 9.2 illustrates the behavior of forward and backward flows between the production system and its market.

The collection of repairable items is also assumed to be a constant percentage, β, of the demand rate, $\beta r = -\beta K(P - P_0)$, where the ratio of Equation 9.27 to Equation 9.1 is presented as $\beta = u(t)/r(t)$. These two assumptions are consistent with the model of Richter (1996a, 1996b), and therefore are adopted here. Richter (1996a, 1996b) assumed that in an interval of length T there are m repair (remanufacturing) batches of size R each, and n production batches of size Q each, where $x = mR + nQ$, $T = x/r(t) = x/r = x/ - K(P - P_0)$. The unit time cost function was given as

$$\psi(n,m,x) = \frac{(nS_P + mS_R)r}{x} + h_f \frac{x}{2}\left[\frac{\alpha^2}{n} + \frac{\beta^2}{m}\right]$$

$$+ h_u \frac{\beta x}{2} + h_u \frac{\beta^2 Q(m-1)}{2m} + (1+\beta)\frac{Er}{x} \tag{9.28}$$

where $E(T) = E = P_0 P/(P_0 - P)$ and $r = -K(P - P_0)$. The optimal lot size that minimizes Equation 9.28 is determined by setting $\partial \psi/\partial x = 0$ and solving for x in terms of the other variables and parameters to get

$$x(n,m) = \sqrt{\frac{2r(nS_P + mS_R + (1+\beta)E)}{h_f\left(\frac{\alpha^2}{n} + \frac{\beta^2}{m}\right) + h_u\beta\left(1 + \beta\frac{m-1}{m}\right)}} \tag{9.29}$$

Equation 9.28 is also convex over n and m. Equations 9.27 and 9.28 were solved using a four-step solution procedure. The results of Jaber and Rosen (2008) suggested that it is cheaper per unit cost to control the flow of an item in the forward flow (from the production system to the market), or in the backward flow (collection of used items from the market for recovery by the production system), when the lot size (production + recovery) is large rather than small. Their results also suggested that accounting for entropy costs may be more relevant for low-demand and expensive items, than for high-demand and inexpensive items.

9.6 A PRICE-QUALITY DRIVEN ECONOMIC ORDER QUANTITY MODEL

The work of Jaber et al. (2004) assumed that a commodity flows from the production/inventory system to the market as a result of a price differential between the product's selling price and its market price, while the quality differential remains unchanged. That is, and as in Equation 9.1, demand flows as a result of reduction in price only. This section postulates that commodity flow is driven by both price and quality.

There is a general consensus among researchers that the demand for a product is dependent on price and quality, where demand increases as the price (quality index) of a product decreases and vice versa. Some of these studies are those of Dorfman and Steiner (1954), Cowling and Rayner (1970), Spence (1975), Kalish (1983), Narasimhan et al. (1993), Teng and Thompson (1996), and Vörös (2002). Therefore and without loss of generality, the demand function in Equation 9.1 can then be written as

$$r(t) = -\kappa \left[\frac{P(t)}{g(t)} - \frac{P_0(t)}{g_0(t)} \right] \qquad (9.30)$$

where $P(t)$ and $g(t)$ $(0 < g(t) < 1)$ are the price and the quality index of the product, respectively, at time t. For the flow to occur, $P(t)/g(t) < P_0(t)/g_0(t)$ for every $t > 0$.

The EOQ model assumes a constant demand rate, that is, $r = r(t)$ where $t \in [0,\infty)$, which implies that Equation 9.1 reduces to $r = -\kappa(P/g - P_0/g_0)$, such that $P/g < P_0/g_0$. Therefore, the entropy cost per cycle given in Equation 9.11 could then be written as

$$E(T) = \frac{\int_0^T -\kappa(P/g - P_0/g_0)dt}{\kappa \int_0^T \left(\frac{P/g}{P_0/g_0} + \frac{P_0/g_0}{P/g} - 2 \right) dt} = -\frac{PP_0}{g_0 P - P_0 g} \qquad (9.31)$$

For Equation 9.31 to be positive $g_0 P < P_0 g \Rightarrow g_0/g < P_0/P$. The total cost per unit of time is developed in a similar manner to Equation 9.12 after replacing Equations 9.1 and 9.5 with Equations 9.30 and 9.31, respectively, and is given as

$$\psi(T) = \frac{A}{T} - h\kappa(P/g - P_0/g_0)\frac{T}{2} - \frac{PP_0}{(g_0 P - g P_0)T}$$

$$- m\kappa(P/g - P_0/g_0) + \gamma \frac{g - g_0}{g_0 T} \qquad (9.32)$$

where $g_0P - gP_0 < 0 \Rightarrow g_0P < gP_0 \Rightarrow g_0/g < P_0/P$, m is the unit purchasing cost ($m < P$), and γ is the cost of improving the quality of the product by 1 percent above the market quality (g_0) of products made by the competitors. Here it is assumed that the initial product quality index is at least equal to that of its competitors. The profit per unit of time is given as

$$\varphi(T) = -(P-m)\kappa(P/g - P_0/g_0) - \frac{A}{T} + h\kappa(P/g - P_0/g_0)\frac{T}{2}$$

$$+ \frac{PP_0}{(g_0P - gP_0)T} - \gamma\frac{g - g_0}{g_0T} \tag{9.33}$$

The optimal cycle time is given by setting the first derivative of Equation 9.33 equal to zero and solving for T to get

$$T^* = \sqrt{\frac{2\left(A - \frac{PP_0}{g_0P - gP_0} + \gamma\frac{g - g_0}{g_0}\right)}{-h\kappa\left(P/g - P_0/g_0\right)}} \tag{9.34}$$

Substituting Equation 9.34 in Equation 9.33 reduces it to

$$\varphi(P,g) = -(P-m)\kappa(P/g - P_0/g_0)$$

$$- \sqrt{-2h\kappa(P/g - P_0/g_0)\left(A - \frac{PP_0}{g_0P - gP_0} + \gamma\frac{g - g_0}{g_0}\right)} \tag{9.35}$$

The optimal solution (P^*, g^*) for Equation 9.35 can be determined by solving the following nonlinear programming problem (NLPP)

Maximize

$$Z = \varphi(P,g) \tag{9.36a}$$

subject to

$$m \leq P \leq P_0 \tag{9.36b}$$

$$g_0 \leq g \leq 1 \tag{9.36c}$$

where P and $g > 0$.

To illustrate, consider an inventory situation with the following input parameters: $\kappa = 20$, $P_0 = 105$, $m = 50$, $g_0 = 0.7$, $h = 1$, $A = 200$, and $\gamma = 50{,}000$. Substituting these in Equation 9.35 and optimizing the NLPP in Equations 9.36a–9.36c, the optimal solution is attained at $g^* = 1$ and $P^* = 101.65$, where $\varphi(P^*, g^*) = 43{,}431$ and $T = 6.737$ ($Q = T \times r = 6.737 \times 967 = 6515$). For this case, the cost to control the flow of a single unit from the inventory system to the market (unit entropy cost) is given from Equation 9.31 as $E/Q = 0.05$. This could be described as an aggressive product diffusion strategy where the firm competes on both quality ($g^* = 1 > g_0 = 0.7$) and price ($P^* = 101.65 < P_0 = 105$). Now, assume that γ is doubled ($\gamma = 100000$), then the optimal solution is attained at $g^* = 0.7$ and $P^* = 77.53$ where $\varphi(P^*, g^*) = 20{,}618$ and $T = 1.26$ ($Q = 1.26 \times 785 = 989$), with $E/Q = 0.43$. This is a strategy where the firm has the competitive advantage to compete on price alone ($g^* = g_0 = 0.7$ and $P^* = 77.53 < P_0 = 105$), since it is too expensive to improve the product quality index beyond its initial value $g = g_0 = 0.7$. The aggressive strategy suggests ordering in larger lots less frequently as it is cheaper to control the flow of the product ($0.05 < 0.43$) from the firm's inventory system to its market, and it is more profitable ($43{,}431 > 20{,}618$).

Suppose now $m = 85$; for the same values of the input parameters suggested in the first example, the optimal solution is attained at $g^* = 1$ and $P^* = 119.57$, where $\varphi(P^*, g^*) = 15{,}839$ and $T = 8.544$ ($Q = T \times r = 8.544 \times 608.70 = 5201$). For this case, the cost to control the flow of a unit from the system to the market is $E/Q = 0.113$. This strategy suggests that the firm has the advantage to compete on the quality index alone ($g^* = 1 > g_0 = 0.7$ and $P^* = 119.57 > P_0 = 105$).

9.7 SUMMARY AND CONCLUSIONS

This chapter presented a survey of the recent works that models commodity flow as an energy flow closing boundaries (system to market). These surveyed works postulated that the behavior of production systems resembles that of physical systems. Such a parallel suggests that production systems may be improved by applying the first and second laws of thermodynamics to reduce system entropy (disorder). The rationale for this new concept is to account for the hidden, difficult-to-estimate costs that usually are not accounted for when analyzing inventory systems.

This entropy (disorder) cost (Jaber et al. 2004) was investigated in different contexts including inventory management (Jaber et al. 2004, 2009; Jaber 2007, 2009, 2009a, 2009b,), supply chain management (Jaber et al. 2006), and reverse logistics (Jaber and Rosen 2008). These studies suggested that when accounting for entropy (disorder) costs, it is cheaper to manage large lots than the small ones. It was further suggested that it might be more relevant to account for entropy for expensive items than for inexpensive ones.

This chapter also extended the model of Jaber et al. (2004) assuming that a commodity flows from the production/inventory system to the market as a result of price and quality differentials, rather than price alone. A basic mathematical model was developed with illustrative numerical examples. The numerical results described the conditions for three commodity flow strategies. The first strategy, an aggressive one, proposed that a firm competes in its market on both price and quality. This is possible when investing to improve quality is economically feasible and when the unit cost is relatively low compared to price. The second strategy proposed that a firm competes on price alone when it is too expensive to improve the quality of the product. The third and last strategy proposed that a firm competes on quality alone when the unit cost is relatively high compared to price. The models surveyed in this chapter could be revisited and investigated for both price and quality. This is left for future work.

REFERENCES

Adkins, A. C. 1984. EOQ in the real world. *Prod Inventory Manag* 25(4):50–54.

Arrow, K. J., and G. Debreu. 1954. Existence of an equilibrium for competitive economy. *Econometrica* 22(3):265–290.

Banerjee, A. 1986. A joint economic lot size model for purchaser and vendor. *Decis Sci J* 17(3):292–311.

Barish, N. N. 1963. Operations research and industrial engineering: The applied science and its engineering. *Oper Res* 11(3):387–398.

Beamon, B. M. 1999. Designing the green supply chain. *Logist Inform Manag* 12(4): 332–342.

Bonney, M. 1994. Trends in inventory management. *Int J Prod Econ* 35(1–3):107–114.

Bonney, M., S. Ratchev, and I. Moualek. 2003. The changing relationship between production and inventory examined in a concurrent engineering context. *Int J Prod Econ* 81–82:243–254.

Callioni, G., X. de Montgros, R. Slagmulder, L. N. van Wassenhove, and L. Wright. 2005. Inventory-driven costs. *Harv Bus Rev* 83(3):135–142.

Cavinato, J. 1991. Lowering set-up costs. *Chilton's Distrib* 90(18):52–53.

Cengel, Y., and M. Boles. 2002. *Thermodynamics: An engineering approach*. Boston: McGraw-Hill.

Chen, W.-H. 1999. Business process management: A thermodynamics perspective. *J Appl Manag Stud* 8(2):241–257.

Chikán, A. 2007. The new role of inventories in business: Real world changes and research consequences. *Int J Prod Econ* 108(1–2):54–62.

Cowling, K., and A. J. Rayner. 1970. Price, quality, and market share. *J Polit Econ* 78(6): 1292–1309.

Crawford, K. M., J. H. Blackstone, and J. F. Cox. 1988. A study of JIT implementation and operating problems. *Int J Prod Res* 26(9):1561–1568.

Crowther, J. 1964. Rationale of quantity discounts. *Harv Bus Rev* 42(2):121–127.

Crusoe, J., G. Schmelzle, and T. Buttross. 1999. Auditing JIT implementation. *Intern Audit* 14(4):21–24.

Drechsler, F.S. 1965. Management and the laws of disorder. *Sci Bus* (winter issue):287.

Drechsler, F. S. 1968. Decision trees and the second law. *Oper Res Q* 19(4):409–419.

Dolan, R. J. 1987. Quantity discounts: Managerial issues and research opportunities. *Market Sci* 6(1):1–22.

Dorfman, R., and P. O. Steiner. 1954. Optimal advertising and optimal quality. *Am Econ Rev* 44(5):826–836.

Fiksel, J. 1996. *Design for environment: Creating eco-efficient products and processes*. New York: McGraw-Hill.

Fisher, H., and D. Siburg. January 2003. A publisher's cash management plan Part 4—Managing your inventory. *PMA Newsletter*, 1–5.

Fleischmann, M., J. M. Bloemhof-Ruwaard, R. Dekker, E. van der Laan, J. A. E. E. van Nunen, and L. N. van Wassenhove. 1997. Quantitative models for reverse logistics: A review. *Eur J Oper Res* 103(1):1–17.

Goodeve, C. 1953. Operational research as a science. *J Oper Res Soc Am* 1(4):166–180.

Gooley, T. B. 1995. Finding the hidden cost of logistics. *Traffic Manag* 34(3):47–50.

Goyal, S. K. 1977. An integrated inventory model for a single supplier-single customer problem. *Int J Prod Res* 15(1):107–111.

Goyal, S. K. 1985. Economic order quantity under conditions of permissible delay in payments. *J Oper Res Soc* 36(4):335–338.

Goyal, S. K., and B. C. Giri. 2003. The production–inventory problem of a product with time varying demand, production and deterioration rates. *Eur J Oper Res* 147(3):549–557.

Gungor, A., and S. M. Gupta. 1999. Issues in environmentally conscious manufacturing and product recovery: A survey. *Comput Ind Eng* 36(4):811–853.

Haley, C. W., and R. C. Higgins. 1973. Inventory policy and trade credit financing. *Manag Sci* 20(4):464–471.

Harris, F. W. 1913. How many parts to make at once? *Factory: The Magazine of Management* 10(2):136–152. (Reprinted in *Oper Res* 38(6):947–950, 1990.)

Harris, T. 1999. The second law of supply chain management. *IIE Solutions* 31(3):24.

Jaber, M. Y. 2007. Lot sizing with permissible delay in payments and entropy cost. *Comput Ind Eng* 52(1):78–88.

Jaber, M. Y., and M. Bonney. 1998. The effects of learning and forgetting on the optimal lot size quantity of intermittent production runs. *Prod Plann Contr* 9(1):20–27.

Jaber, M. Y., M. Bonney, and I. Moualek. 2009a. Lot sizing with learning, forgetting and entropy cost. *Int J Prod Econ*, doi: 10.1016/j.ijpe.2008.08.006.

Jaber, M. Y., M. Bonney, and I. Moualek. 2009b. An economic order quantity model for an imperfect production process with entropy cost. *Int J Prod Econ*, doi: 10.1016/j. ijpe.2008.08.007. 118(1):26–33.

Jaber, M. Y., M. Bonney, M. A. Rosen, and I. Moualek. 2009. Entropic order quantity (EnOQ) model for deteriorating items. *Appl Math Model* 33(1):564–578.

Jaber, M. Y., R. Y. Nuwayhid, and M. A. Rosen. 2004. Price-driven economic order systems from a thermodynamic point of view. *Int J Prod Res* 42(24):5167–5184.

Jaber, M. Y., R. Y. Nuwayhid, and M. A. Rosen. 2006. A thermodynamic approach to modelling the economic order quantity. *Appl Math Model* 30(9):867–883.

Jaber, M.Y. and M.A. Rosen 2008. The economic order quantity repair and waste disposal model with entropy cost. *Eur J Oper Res* 188(1):109–120.

Jones, D. J. 1991. JIT & the EOQ model: Odd couple no more! *Manag Account* 72(8):54–57.

Kalish, S. 1983. Monopolist pricing with dynamic demand and production cost. *Market Sci* 2(2):135–159.

Karp, A., and B. Ronen. 1992. Improving shop floor control: An entropy model approach. *Int J Prod Res* 30(4):923–938.

Kingsman, B. G. 1983. The effect of payment rules on ordering and stockholding in purchasing. *J Oper Res Soc* 34(11):1085–1098.

Konar, S. and M. A. Cohen. 2001. Does the market value environmental performance? *Rev Econ Stat* 83(2):281–289.

Lummus, R. R. 1995. Are confusion costs stealing your profits? *Ind Manag* 37(3):26–28.

McCarthy, I. P., T. Rakotobe-Joel, and G. Frizelle. 2000. Complex systems theory: Implications and promises for manufacturing organisations. *Int J Manuf Tech Manag* 2(1–7): 559–579.

Narasimhan, R., S. Ghosh, and D. Mendez. 1993. A dynamic model of product quality and pricing decisions on sales responses. *Decis Sci* 24(5):893–908.

Osteryoung, J. S., E. Nosari, D. E. McCarty, and W. J. Reinhart. 1986. Use of the EOQ model for inventory analysis. *Prod Inventory Manag* 27(3), 39–46.

Pendlebury, J., and R. Platford. 1988. The heavy hidden cost of materials handling. *J Cost Anal Manag* 2(1):4–8.

Porteus, E. L. 1986. Optimal lot sizing, process quality improvement, and setup cost reduction. *Oper Res* 34(1):137–144.

Rafaat, F. 1991. Survey of literature on continuously deteriorating inventory models. *J Oper Res Soc* 42(1):27–37.

Richter, K. 1996a. The EOQ and waste disposal model with variable setup numbers. *Eur J Oper Res* 95(2):313–324.

Richter, K. 1996b. The extended EOQ repair and waste disposal model. *Int J Prod Econ* 45 (1–3):443–447.

Ronen, B., and R. Karp. 1994. An information entropy approach to the small-lot concept. *IEEE Trans Eng Manag* 41(1):89–92.

Salameh, M. K., N. E. Abboud, A. N. El-Kassar, and R. E. Ghattas. 2003. Continuous review inventory model with delay in payments. *Int J Prod Econ* 85(1):91–95.

Samaras, T. T. 1973. Overcome entropy or your organization dies. *Office* 77(3):13.

Selen, W. J., and W. R. Wood. 1987. Inventory cost definition in an EOQ model application. *Prod Inventory Manag* 28(4):44–47.

Shannon, C. E. 1948. A mathematical theory of communication. *Bell Syst Tech J* 27(3):379–423.

Spence, A. M. 1975. Monopoly, quality, and regulation. *Bell J Econ* 6(2):417–429.

Sprague, L. G. 2002. Inventory management. In *International encyclopedia of business & management–manufacturing & operations*, 2nd ed., edited by L. Sprague, 4849–4954. London: Thomson International Press.

Teng, J., and G. L. Thompson 1996. Optimal strategies for general price-quality decision models of new products with learning production costs. *Eur J Oper Res* 93(3):476–489.

Tseng, K.-J. 2004. Application of thermodynamics on product life cycle. *J Am Acad Bus Camb* 4(1/2):464–470.

Tyler, G. W. 1989. A thermodynamic model of manpower systems. *J Oper Res Soc* 40(2): 137–139.

Ullmann, J. E. 1982. White-collar productivity and growth of administrative overhead. *Natl Prod Rev* 1(3):290–300.

Uzawa, H. 1960. Walras' *tâtonnement* in the theory of exchange. *Rev Econ Stud* 27(3): 182–194.

Vörös, J. 2002. Product balancing under conditions of quality inflation, cost pressures and growth strategies. *Eur J Oper Res* 141(1):153–166.

Waters, D. 2003. *Inventory control and management*. London: Wiley.

Whewell, R. 1997. Turning up the heat under the supply-chain. *Logist Focus J Inst Logist* 5(4):18–22.

Wilson, A. G. 1970. The use of the concept of entropy in system modelling. *Oper Res Q* 21(2):247–265.

Woolsey, G. 1988. A requiem to the EOQ. *Prod Inventory Manag* 29(3):44–47. (Reprinted in *Hosp Mater Manag Q* 12(1):82–90, 1990)

Index

Milton Keynes UK
Ingram Content Group UK Ltd.
UKHW040104071024
449327UK00019B/794